Earth and Environmental Sciences Library

Earth and Environmental Sciences Library (EESL) is a multidisciplinary book series focusing on innovative approaches and solid reviews to strengthen the role of the Earth and Environmental Sciences communities, while also providing sound guidance for stakeholders, decision-makers, policymakers, international organizations, and NGOs.

Topics of interest include oceanography, the marine environment, atmospheric sciences, hydrology and soil sciences, geophysics and geology, agriculture, environmental pollution, remote sensing, climate change, water resources, and natural resources management. In pursuit of these topics, the Earth Sciences and Environmental Sciences communities are invited to share their knowledge and expertise in the form of edited books, monographs, and conference proceedings.

Sk Asraful Alam · Ramkrishna Maiti

Sediment Source to Sink: Deciphering Sediment Connectivity to Large Dams in Damodar River Basin

Springer

Sk Asraful Alam
Department of Geography
Vidyasagar University
Midnapore, West Bengal, India

Ramkrishna Maiti
Department of Geography
Vidyasagar University
Midnapore, West Bengal, India

ISSN 2730-6674 ISSN 2730-6682 (electronic)
Earth and Environmental Sciences Library
ISBN 978-3-031-87071-2 ISBN 978-3-031-87072-9 (eBook)
https://doi.org/10.1007/978-3-031-87072-9

This Springer imprint is published by the registered company Springer Nature Switzerland AG
The registered company address is: Gewerbestrasse 11, 6330 Cham, Switzerland

Preface

The Damodar River Basin, a lifeline for millions, exemplifies the intricate interplay between natural processes and human interventions. This book, *Sediment Source to Sink: Deciphering Sediment Connectivity to Large Dams in Damodar River Basin*, provides a thorough investigation into sediment dynamics, fluvial connectivity, and the challenges posed by soil erosion and sediment deposition in large reservoirs. It begins with foundational concepts, such as fluvial connectivity, soil erosion hotspots, and sediment delivery ratios, while exploring their implications for dam sedimentation and hydrological stability. The Damodar River Basin (DRB) faces escalating challenges from human alterations and environmental encroachment. The intensified erosion in the DRB has resulted in significant sedimentation within reservoirs, reducing their storage capacity and jeopardizing the sustainability of water resources and infrastructures. Tackling this issue demands integrated management strategies that account for both natural processes and human-induced impacts to mitigate sedimentation and ensure the longevity of reservoirs. Through a synthesis of global perspectives and regional insights, the book builds a conceptual framework that bridges theory and practice in the domain of sediment management.

Using advanced modeling techniques like SWAT and RUSLE, the book quantifies sediment yield, water discharge, and surface runoff across multiple temporal and spatial scales. The construction of multiple dams in the upper watershed of the Damodar River Basin has disrupted the basin's overall sediment connectivity, altering the natural flow and sediment transport processes. The increased sediment output and intensified soil erosion in the upper reaches are directly linked to the progressive loss of storage capacity in key reservoirs and dams, including Maithon, Panchet, Tenughat, Konar, and Tilaya. The book highlights the significant impacts of soil erosion hotspots on sedimentation, linking them to land use dynamics and sediment connectivity indices. A detailed analysis of three major dams—Maithon, Panchet, and Tenughat—unravels the source-to-sink sediment relationships, offering critical insights into basin-scale dis-connectivity and sediment pathways over time. Furthermore, the exploration of land use and land cover changes reveals the anthropogenic influences on sediment variation, emphasizing the pressing need for sustainable land and water management practices. This highlights the need for integrated sediment management strategies to mitigate reservoir capacity loss and ensure the sustainable functioning of the basin's hydrological systems.

This book not only presents groundbreaking findings but also offers actionable management strategies and recommendations for mitigating sedimentation challenges in dam catchments. It also examines sediment dynamics, fluvial discharge, surface runoff, and sediment connectivity index (SCI). The coupling of SEH, SCI, and soil erosion probability (SE) has been used to identify the major sub-basins contributing to sediment transport. It will become a valuable resource for researchers, policymakers, and practitioners in hydrology, geomorphology, and watershed management. Addressing the sedimentation problem through a multidisciplinary lens paves the way for innovative solutions to enhance the sustainability of large dams and water resources. We hope that this work will inspire further research and effective interventions to tackle sedimentation issues in similar hydrological settings worldwide.

Midnapore, West Bengal, India

Sk Asraful Alam
Ramkrishna Maiti

Acknowledgments The authors wish to acknowledge the invaluable support and assistance of the colleagues and co-reseracher s in the Department of Geography, Vidyasagar University, Medinipur, India, in this research work.

We wish to dedicate this work to the memory of Dr. Moumita Moitra (Maiti), whose unwavering encouragement, profound humility, and inspiring dedication to research ignited our passion in this field. Her motivation was an invaluable source of strength and inspiration throughout this journey.

We would also like to express our sincere gratitude to all the officials from various departments, including the Indian Meteorological Department (IMD, Pune), Water Resources Information System of India (WRIS), Central Water Commission (CWC), Damodar Valley Corporation (DVC), Food and Agricultural Organization (FAO), and the Irrigation and Waterways Department, Government of West Bengal, for providing us with valuable information related to our study area.

We gratefully acknowledge the invaluable help and support provided by Dr. Abhisek Ghosh, Dr. Suman Bera, Dr. Souvanik Maity, Dr. Swapan Kumar Maiti, Dr. Animesh Majee, Dr. Prasenjit Bhunia, Dr. Sonam Lama, Dr. Surojit Mondal, Dr. Hirok Kumar Mahata, Dr. Anwesha Banerjee, Dr. Pritha Das, and all of my co-researchers, whose contributions have been instrumental in various aspects of this work. We would like to sincerely acknowledge Farhina Akhtar for her invaluable assistance during the field survey for our research. Her support and collaboration played a vital role in completing this part of this book.

We are deeply grateful for the insightful guidance of Prof. Sujit Mandal, Dr. Prasenjit Acharya, and Dr. Sikhendra Kishor De, who generously dedicated their valuable time whenever we sought their advice and provided us with direction and support.

Last, but certainly not least, our acknowledgment would be incomplete without expressing our heartfelt gratitude to our family, the greatest source of our strength. Their unwavering encouragement, cooperation, and continuous moral and emotional support have been invaluable. This work would not have been possible without their selfless love and constant support at every step.

Midnapore, West Bengal, India

Sk Asraful Alam
Ramkrishna Maiti

Competing Interests The authors have no competing interests to declare that are relevant to the content of this manuscript.

Contents

About the Authors

Sk Asraful Alam is a Senior Research Fellow (SRF) in the Department of Geography at Vidyasagar University, Midnapore, West Bengal, India. He earned his M.Phil. degree from Vidyasagar University in the Centre for Earth System Science. He has also served as a guest lecturer in the Department of Geography at Midnapore College, West Bengal, India. He has published research articles in prestigious national and international journals. He is also serving as a reviewer for national and international journals. His research interests primarily focus on geomorphology, with particular emphasis on fluvial geomorphology, river bank erosion, paleochannel and paleoclimatic reconstruction, landscape evolution, geomorphic hazards, climate change impacts, sediment budgets, and fluvial connectivity.

Prof. Ramkrishna Maiti presently serves Vidyasagar University in the post of professor in the Department of Geography, Midnapore, West Bengal, India. His significant contributions have been in the areas of Geomorphology, Hydrology, Environmental Hazards, and Geographical Philosophy. He leads several research projects funded by the UGC and ICSSR. He successfully guided eighteen Ph.D. theses and four M.Phil. theses. He has over 125 publications to his credit in the national and international journals of repute. He authored several books including *Semi-Quantitative Approach for Landslide Assessment and Prediction* (Springer), *Sedimentation in the Rupnarayan River* (2 volumes) (Springer), *Modern Approaches to Fluvial Geomorphology* (Primus Books), *Applied Geomorphology and*

Contemporary Issues (Edited Books, Springer), *Land Acquisition, Industrialization and Livelihoods A case study on JSW Bengal Steel Plant* (Springer) *Development of Geographical Thought* (Nabodaya Publishers, Kolkata), and *Geotectonics and Geomorphology* (Nabodaya Publishers, Kolkata).

Abbreviations

ANOVA	Analysis of Variance
BDI	Basin Dysconnectivity Index
CFRG	Coarse Fragment Factor
CGWB	Central Ground Water Board
CL	Confidence Level
CREAMS	Chemical, Runoff and Erosion from Agricultural Management Systems
CWC	Central Water Commission
DD	Drainage Density
DRB	Damodar River Basin
DSMW	Digital Soil Map of the World
DVC	Damodar Valley Corporation
EPIC	Erosion—Productivity Impact Calculator
ESRI	Environmental Systems Research Institute
ET	Evapotranspiration
EUROSEM	European Soil Erosion Model
FAO	Food and Agricultural Organization
FTR	Failed to Reject
GLEAMS	Groundwater Loading Effects on Agricultural Management Systems
HRUs	Hydrological Response Units
IC	Index of Connectivity
ICOLD	International Commission on Large Dams
ILSM	Integrated agricultural and Soil Management
IMD	Indian Meteorological Department
IUCN	International Union for Conservation of Nature
IWMP	Integrated Watershed Management Plan
LS	Slope Length Factor
LULC	Land Use Land Cover
MSL	Mean Sea Level
MUSLE	Modified Universal Soil Loss Equation
NASA	National Aeronautics and Space Administration

NDVI	Normalized Difference Vegetation Index
NRCS	Natural Resource Conservation Service
RBO	River Basin Organization
RI	Roughness Index
ROTO	Routing Outputs and Outlet
RR	Relative Relief
RUSLE	Revised Universal Soil Loss Equation
SB	Sub-Basin
SCI	Sediment Connectivity Index
SCS	Soil Conservation Service
SD	Standard Deviation
SDR	Sediment Delivery Ratio
SE	Soil Erosion
SEDD	Sediment Distributed Delivery
SEH	Soil Erosion Hotspot
SRTM	Shuttle Rader Thematic Mapping
SWAT	Soil and Water Assessment Tool
SWRRB	Simulator for Water Research Basins
SY	Sediment Yield
UNCDD	United Nations Convention to Combat Desertification
USDA	United State Department of Agriculture
USLE	Universal Soil Loss Equation
WCD	World Commission on Dams
WEPP	Water Erosion Prediction Project

Symbols

a_i	Surface roughness coefficient (n)
$C_{i,j}$	Land coverage factor (Unit less)
E_a	Amount of evapotranspiration on the *ith* day (mm)
$K_{i,j}$	Soil erodibility factor (t ha h ha^{-1} MJ^{-1} mm^{-1})
l_i	Flow length (m) of the cell *i*
$LS_{i,j}$	Slope length factor (Unit less)
$P_{i,j}$	Support practice factor (Unit less)
Q_{gw}	Amount of return flow on the *ith* day (mm)
Q_{peak}	Runoff peak discharge (m^3/s)
Q_{surf}	Amount of surface runoff on the *ith* day (mm)
R_{day}	Amount of rainfall on the *ith* day (mm)
$R_{i,j}$	Rainfall erosivity factor (MJ mm h^{-1} ha^{-1} $year^{-1}$)
S_i	Slope gradient
t_i	Travel time (hours) for cell *i*
Te	Trap efficiency (% in decimal units)

V_i	Flow velocity for the cell (ms^{-1})
V_i	The Inflow sediment load
V_o	The Outflow sediment load
W_{seep}	Water seepage to groundwater from the soil profile on the *ith* day (mm)
β	Basin Specific Parameter
$\sqrt{A}$	Upslope contributing area (m^2)
$\Delta\tau_\Gamma$	The local residence time changes in years

List of Figures

List of Tables

Chapter 1
Introduction

Abstract This chapter delves into sediment dynamics in the Damodar River Basin, focusing on the interplay of fluvial connectivity, soil erosion hotspots, and reservoir sedimentation. The research explores how natural sediment pathways are altered by river regulation and anthropogenic activities, emphasizing the importance of understanding sediment connectivity from source to sink. A comprehensive conceptual framework integrates global and regional perspectives, addressing sediment transport mechanisms and their implications for dam sedimentation. Advanced methodologies, supported by diverse datasets, are employed to identify sediment sources, assess connectivity, and quantify sediment production. The findings have significant implications for watershed management, soil conservation, and sustainable reservoir operation, offering actionable strategies to mitigate sedimentation challenges. This chapter bridges theoretical insights with practical solutions, contributing to the effective management of sediment in regulated river systems.

Keywords Fluvial connectivity · River regulation · Reservoir sedimentation · Sedimentation · Digital elevation model · Valley

1.1 Introduction

Rivers are the great shapers of terrestrial landscapes [95]. Therefore, the rivers continuously shape and reshape the landscape through process-form interactions, a dynamic interplay between hydrological processes and the resulting landforms. A drainage basin is a complete system that collects and channels entire runoff into a single outlet [56], Through systematic interaction involving input, processing, and output. Thus, fluvial geomorphology, analyzed across different spatiotemporal scales with distinct and well-defined topographical units, is referred to as a nested hierarchical system [19, 56, 80]. As a result, smaller drainage basins contribute their discharge to larger drainage basins, which in turn flow into even larger drainage systems [25, 56]. Hence, The entire river system is broadly connected through bidirectional fluxes

S. A. Alam and R. Maiti, *Sediment Source to Sink: Deciphering Sediment Connectivity to Large Dams in Damodar River Basin*, Earth and Environmental Sciences Library,
https://doi.org/10.1007/978-3-031-87072-9_1

of energy and matter between the river network's courses and the surrounding landscape [95]. However, sub-catchments within a drainage basin often display unique physical attributes, influenced by their specific topography, geology, and land cover. So, connectivity plays an important role in understanding the complex pathways and fluxes of material, energy, and organisms in the fluvial system [96]. The theme of the 47th Binghamton Geomorphology Symposium (2017), centered on connectivity in geomorphology, outlined six distinct categories of connectivity: *Sediment connectivity, hydrologic connectivity, geochemical connectivity, riverine connectivity, landscape connectivity*, and *modeling connectivity* [96]. The connectivity responses to geomorphic change vary from catchment to catchment [11]. A catchment operates through the interconnected processes of longitudinal, lateral, and vertical linkages [11, 92], which together regulate the movement and distribution of water, sediment, and energy across the landscape. However, researchers have developed an advanced connectivity index approaches by applying a process-based understanding of landscapes in specific settings: sediment connectivity [7, 9, 13, 14, 21, 41, 44, 60, 66, 79, 101], hydrological connectivity [10, 20, 23, 54, 61, 75, 84, 91, 98, 100], and longitudinal connectivity [12, 22, 55, 72, 97].

Sediment inputs to fluvial channel networks exhibit significant variability across both spatial and temporal scales [16]. Anthropogenic impact on connectivity in fluvial systems has significantly altered the natural flow and dynamics of water, sediment, and ecological processes at both spatial and temporal scales. Human activities directly impact connectivity through land use and land cover (LULC) changes [99], as well as various land use practices that modify sediment delivery from hillslopes and floodplains to channel systems [68]. Land degradation, including deforestation, overgrazing, and unsustainable agricultural practices, accelerates soil erosion by weakening the soil structure and reducing vegetation cover [6]. Therefore, soil erosion in the upper catchment can be linked to the assessment of various forms of connectivity and disconnectivity within a river basin [11]. The upper catchment of the Damodar River basin is significantly affected by anthropogenic activities, including open-cast mining, deforestation, illegal coal mining, and land use transformation. Therefore, this region has experienced severe soil erosion due to large-scale mining activities near the watercourse. Thus, the interplay of fluvial connectivity and soil erosion amplifies sediment yield to the main channel and reservoirs. Sediment yield should be linked to the evaluation of connectivity and (dis)connectivity within the catchment system [11].

The Damodar River basin is experiencing substantial soil erosion [47, 50, 70]. Consequently, this eroded soil produced a huge sediment yield input to the fluvial channel and reservoirs. Reservoir sedimentation is one of the important problems across the world [28, 51]. Syvitski et al. [86] reported that global rivers transport an estimated 2.3 ± 0.6 Gt yr-1 of sediment inland due to soil erosion, with Asian rivers (excluding those in Indonesia) contributing approximately 4.74 ± 0.8 Gt annually. Despite this high sediment transport, the overall flux has decreased by 1.4 ± 0.3 Gt due to sediment retention in reservoirs. In the Damodar River basin, reservoir sedimentation is a critical issue caused by excessive sediment influx from the upper catchment. Land degradation and soil erosion intensity are positively correlated, leading to an increase in sediment yield (SY) within the channel. This sediment accumulation gradually reduces the storage capacity of reservoirs and dams, including Maithon, Panchet, Tenughat, Konar, and Tilaya in the Damodar River basin.

Contrarily, dam decommissioning has become increasingly common in many parts of the world as aging infrastructure, environmental concerns, and changing societal needs drive the removal of dams. This process aims to restore river connectivity, improve ecosystem health, and mitigate sedimentation issues [3, 40, 59, 63, 82]. Excessive sedimentation in reservoirs is a critical issue that may lead to dam decommissioning in the future, as it impacts storage capacity, dam safety, and ecosystem health. So, it is essential to identify and analyze the sediment inflow to reservoirs and the sediment production source areas within their catchments.

The Damodar River Basin (DRB) is heavily impacted by human activities, including encroachment and the construction of multiple dams and reservoirs. Flowing through one of the world's richest mineral belts, the Damodar River basin supports extensive mining and industrial operations. However, these activities have made the basin highly susceptible to severe soil erosion. So, identifying soil erosion zones in the upper catchment and their sediment pathways is essential for understanding reservoir sedimentation and sediment outflow dynamics. Under these conditions, this study employs an integrated approach using RUSLE, IC, SDR, SWAT, and SEH methods to estimate sediment yield (SY), catchment discharge, and surface runoff, while also assessing soil erosion rates (SE), sediment connectivity, and hotspots within the Maithon, Panchet, and Tenughat dam catchments in the Damodar River basin. The combined 'RUSLE-IC-SDR-SWAT-SEH' methodology effectively identifies key sediment sources and vulnerable areas that contribute significantly to sedimentation in both the main channel and reservoirs. This chapter underscores the value of this integrated approach in improving understanding of reservoir sedimentation and promoting sustainable soil and water resource management, offering critical information for informed decision-making.

1.2 Key Concepts

1.2.1 Concept of Fluvial Connectivity

Fluvial connectivity has emerged as a key concept in hydrology and geomorphology in recent years, describing the interactions of water and sediment across different landscapes and within river systems [7, 8, 10, 11, 13, 27]. Three primary types of connectivity [8] or linkages [27] have been identified within fluvial watersheds or systems. Fluvial connectivity refers to the water-mediated transfer of materials, energy, organisms, and sediment along and across riverine systems, encompassing their longitudinal (upstream–downstream), lateral (channel-floodplain), and vertical (surface–subsurface) linkage within the landscape [94]. Longitudinal connectivity or linkage refers to the connection along the river's flow direction from the headwaters to the downstream regions including the tributary-trunk strem relationship [26, 42]. Longitudinal connectivity can be disrupted by dams, barrages, and other in-stream barriers that fragment river systems. Channel-floodplain, or lateral connectivity, is governed by the magnitude and frequency of overbank events, facilitating the exchange of materials and organisms between the floodplain and the channel [26]. Vertical linkages within the system involve interactions between surface and subsurface water and sediment [26]. Vertical linkage is the exchange between surface water, groundwater, and the hyporheic zone which indicates the surface–subsurface interaction and the entrainment of sediment [26]. Fluvial connectivity within a catchment system encompasses the interplay of landscape connectivity (hillslope–channel), hydrological connectivity (upstream–downstream), and sedimentological connectivity (both lateral and longitudinal). Therefore, connectivity is crucial for understanding sediment delivery, landform processes, and their evolution.

1.2.2 Concept of River Regulation

River regulation encompasses a range of human-engineered modifications to control or manipulate river flows. Humans have been damming rivers since ancient times [87] to fulfill their livelihood needs. The first significant dams were built in the arid regions of the Middle East, including Mesopotamia and Egypt [87]. Over time, the regulation and control of rivers have continually expanded, with increasing efforts to manage water flow and resources. This trend has grown steadily as the need for flood control, irrigation, and water supply has intensified. The total number of dams worldwide surpassed 800,000 in 2007 [1]. The register of the International Commission on Large Dams (ICOLD) shows a dramatic increase over the past thirteen years, with the number of large dams rising to approximately 62,000 [43]. Therefore, river regulation has significant effects on downstream hydrology, geomorphology, sediment flux, and ecosystems [38, 62, 67, 83, 85, 87, 89, 90, 93]. Thus, river regulation significantly affects the natural connectivity and functioning of river systems. The various dimensions of river connectivity—longitudinal, lateral, and vertical—are altered, leading to disruptions in hydrological, ecological, and sediment dynamics. By modifying the natural flow of water, altering sediment transport, and disrupting the exchange between the river and surrounding ecosystems, regulation transforms the river's ecological and hydrological functions. While river regulation can provide significant benefits (e.g., water supply, flood control, hydropower), these benefits come with trade-offs that need to be carefully managed to minimize negative impacts on river ecosystems and connectivity.

1.2.3 Concept of Soil Erosion Hotspot

A soil erosion hotspot refers to a specific area or region where soil erosion is occurring at a significantly higher rate than the surrounding areas, often resulting in degradation of the landscape [6], loss of soil fertility, and increased sedimentation in water bodies. These hotspots are particularly vulnerable to soil erosion processes, including water and wind erosion, due to a combination of physical, environmental, and human-induced factors. Soil erosion hotspots are often characterized by disturbed or vulnerable landscapes that experience intense natural forces or are heavily impacted by human activities. The Getis-Ord statistic [29] is a spatial statistical method used to identify clusters of high or low values in geographic data, particularly useful for identifying "hotspots" or "cold spots" of certain phenomena. In the context of soil erosion, the Getis-Ord statistic can be used to identify areas where erosion is statistically significant ($p < 0.05$) compared to the surrounding areas ($p > 0.05$), helping to highlight soil erosion hotspots [2, 30, 39, 53]. This method is particularly valuable for identifying soil erosion hotspots, which are considered sediment production zones in the Damodar River Basin. It can be effectively applied to target soil conservation efforts, land-use planning, and management strategies aimed at mitigating soil erosion in the region.

1.2.4 Concept of Reservoir Sedimentation

Reservoir sedimentation refers to the process by which sediment, primarily composed of soil particles such as sand, silt, and clay, accumulates in a reservoir over time. This process occurs when sediment-laden water enters the reservoir, and the water's velocity decreases, causing the sediment to settle out and accumulate at the bottom. Reservoir sedimentation can have significant impacts on the functional storage capacity, water quality, and ecological functions of a reservoir due to sediment accumulation and deposition [24, 65]. Sedimentation is a complex phenomenon [5], which can take place upstream, downstream, and within the reservoir. Soil erosion is typically the primary cause of reservoir sedimentation [18, 64]. Sediment yield has a significant impact on reservoir sedimentation [69] in the Damodar River Basin, contributing to the gradual reduction of reservoir storage capacity, degradation of water quality, and operational inefficiencies in the dams, particularly in reservoirs associated with the Maithon, Panchet, and Tenughat dams [18]. Understanding how sediment yield affects reservoir sedimentation is crucial for managing water storage, optimizing dam operations, and preserving ecological health in the region. By addressing the root causes of high sediment yield and sedimentation, it is possible to preserve the long-term viability of these reservoirs for water storage, power generation, and ecological health.

1.3 Conceptual Framework

The conceptual framework of this study is designed to understand the sediment dynamics and its impact on reservoir sedimentation in the Damodar River Basin (Fig. 1.1). The first step is to identify soil erosion hotspot zones in the upper catchment, which are primarily influenced by anthropogenic activities such as mining, deforestation, and agriculture. These zones are the primary sediment production areas. The second step involves identifying sediment connectivity pathways, which trace how sediment moves from these hotspots to the downstream reservoirs. Next, the book aims to understand the variations in sediment yield at the dam catchment outlets over time, assessing the seasonal and long-term changes in sediment transport. The fourth step is to analyze the linkage between erosional hotspots and sediment connectivity, as the strength of this connection directly affects the amount of sediment transported to the reservoir. The fifth step examines land use and land cover variations and how these factors influence soil erosion rates and sediment yield. Finally, the book seeks to understand the interlinkage between hotspots, connectivity, and sediment deposition in the reservoir, with the hypothesis that strong connectivity will contribute a higher amount of sediment, while poor connectivity will result in reduced sediment deposition, leading to better or worse reservoir management outcomes. The framework used in this book aims to understand the spatial variability of sediment dynamics and propose strategies to mitigate sedimentation, ensuring sustainable water and reservoir management.

1.4 Objectives of the Book

The primary objective of this book is to investigate reservoir sedimentation issues arising from sediment production in the upper catchment and to analyze the associated sediment pathways. The objectives of this book are as follows:

1. Analyze the spatio-temporal variations in sediment yield rate, water discharge, and surface runoff within the Damodar River Basin (DRB).
2. Quantify the variations of soil erosion rate (SE) and sediment delivery ratio (SDR) across the catchment.
3. Assess the sediment connectivity index (IC) in the Damodar catchment to understand sediment transport dynamics.
4. Identify sediment production zones and examine the relationships between sediment yield (SY), sediment delivery ratio (SDR), and soil erosion rate (SE).
5. To investigate the linkage between sediment connectivity and soil erosion hotspots (SEH) within major dam catchments.
6. Understand the impact of land use and land cover dynamics on sediment yield in the channel.

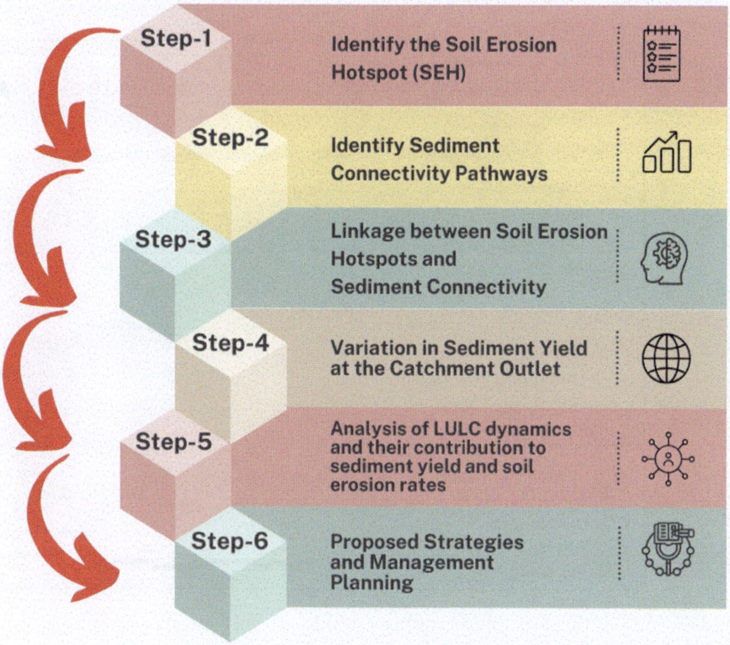

Fig. 1.1 A conceptual framework for studying the Damodar River Basin

1.5 Location of Study Area

The Damodar River Basin (DRB), a sub-watershed of the Ganga River System in India, is a highly rain-fed, shallow, wide, and seasonal tropical river with a fluctuating flow regime. The basin spans across the states of Jharkhand and West Bengal. In Jharkhand, it covers the districts of Chatra, Latehar, Lohardaga, Ranchi, Hazaribagh, Koderma, Giridih, Bokaro, Dhanbad, and Jamtara, while in West Bengal, it includes the districts of Purulia, Bankura, Burdwan, Hugli, and Howrah. The catchment area of the DRB extends from 22°13′ N to 24°30′ N latitude and 84°40′ E to 88°15′ E longitude. The river originates from Khamarpath Hill in the Palamau region of the Chotanagpur Plateau and flows for a distance of 540 km [74]. The main tributaries of the Damodar River include the Barakar, Bokaro, Jamunia, Konar, Behra, and Khadia rivers. The Barakar River flows through Koderma, Hazaribagh, Giridih, and Jamtara districts, originating from the Koderma Plateau and merging with the Damodar River near Dishergarh. Below this confluence, the Nunia and Sali rivers further contribute to the flow. The Damodar River bifurcates near Jamalpur in Bardhaman district into two branches before joining the Bhagirathi-Hugli River. The eastern branch, known as the Amta Channel, joins the Hugli River at Gorchumbak near Ulberia, while the western branch, called the Mundeswari, receives the Dwarkeswar and Siali rivers as right-bank tributaries and ultimately discharges into the Hugli River through the

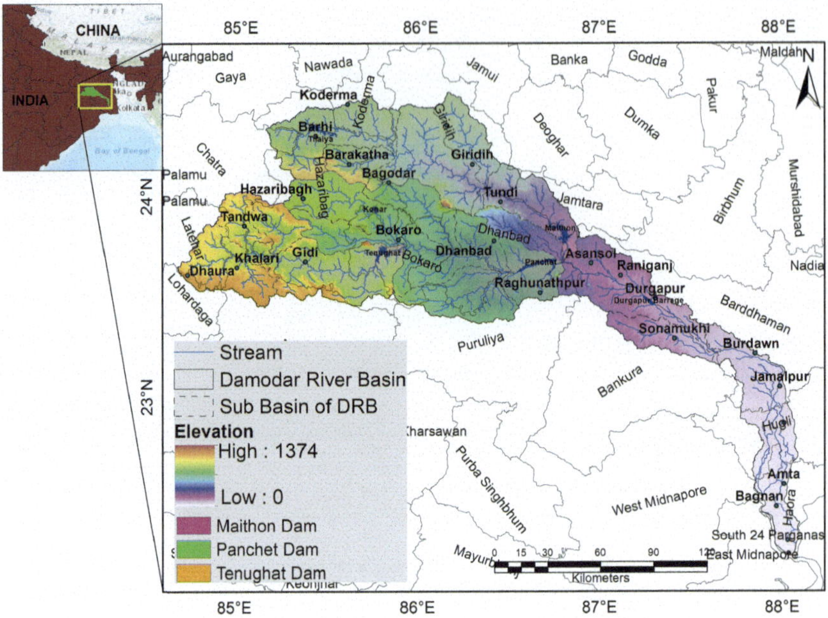

Fig. 1.2 Location map of the study area under investigation in this book

Rupnarayan Estuary [74]. Observations indicate that about three-fourths of the DRB's area lies in Jharkhand, while one-fourth is in West Bengal (Fig. 1.2). The DRB has been divided into 23 sub-basins using the Arc-hydro tool in ArcGIS 10.3 software (https://www.esri.com/en-us/arcgis/geospatial-platform/overview).

1.6 Applied Methodology

Research methodology is a logical approach to guide similar scientific investigation. It involves specific procedures or techniques to identify, select, process, and analyze data. The methodological framework for this research is visually represented in a flow chart (Fig. 1.3), outlining the sequential steps undertaken. This data was sourced from a variety of credible outlets, including published and unpublished works, government offices, e-government portals, and organizational records. These secondary sources provided a robust foundation for the study, ensuring the accuracy and reliability of findings in the absence of primary data.

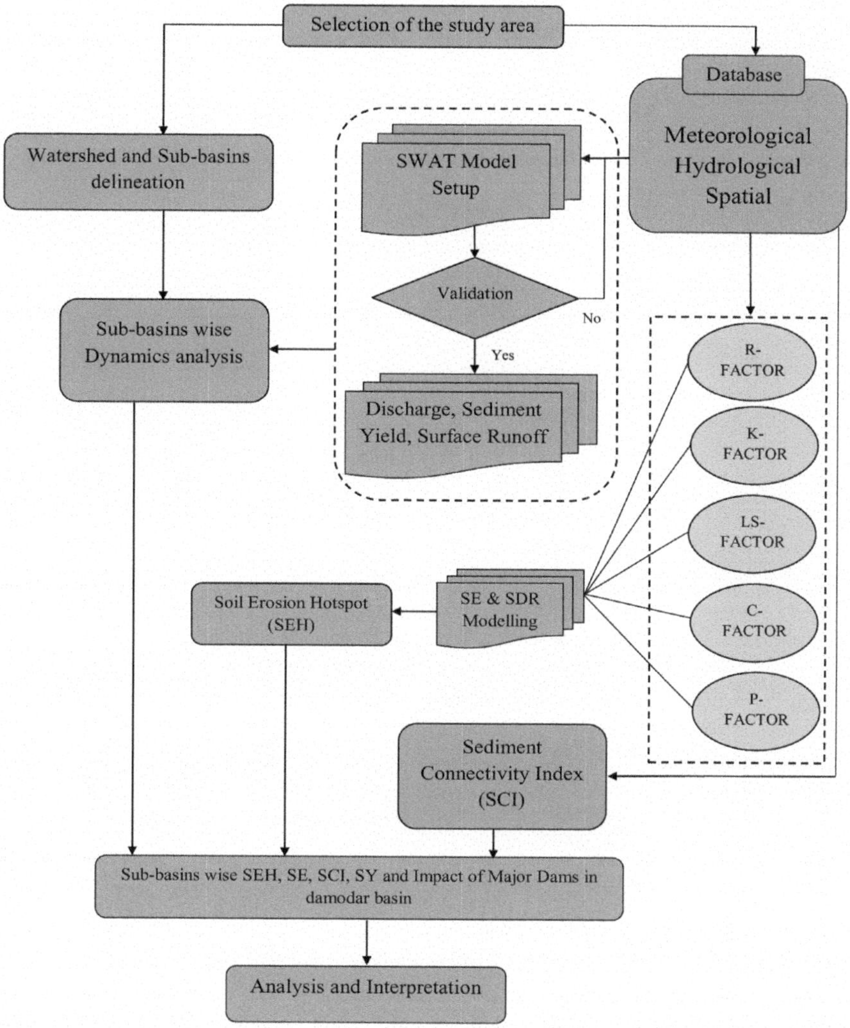

Fig. 1.3 The flowchart outlining the methodology used in this book and applied to the study area and similar studies

1.7 Dataset

The data for this study have been sourced from a variety of reliable platforms to ensure comprehensive coverage of the research objectives (Table 1.1). An extensive field survey was conducted, along with visits to government officials, to collect hydrological and sedimentological data for various dams and reservoirs. These secondary data were gathered from government agencies, published reports, academic studies, and

Table 1.1 Data source of the present study

Feature layer	Source	Thematic data layer	Resolution
Geological map	https://bhukosh.gsi.gov.in	Geological map	1:250,000
Soil map	https://bhukosh.gsi.gov.in https://www.fao.org	Soil map	1:250,000
Geomorphology	https://bhukosh.gsi.gov.in	Geomorphological map	1:250,000
LANDSAT V LANDSAT 8 OLI	https://www.earthexplorer.usgs.gov	NDVI map	30 * 30 m 30 * 30 m
Rainfall data	http://imdpune.gov.in	Rainfall Distribution map	0.25° * 0.25°
DEM	https://www.earthexplorer.usgs.gov	Elevation map, slope map, drainage density, relative relief	30 * 30 m
Land use land cover (LULC) map	https://daac.ornl.gov/cgi-bin/dsviewer.pl?ds_id=1336 https://livingatlas.arcgis.com/landcover/	Land use land cover map of 1990, 2005, and 2021	100 * 100 m 100 * 100 m 10 * 10 m
Weather data	https://globalweather.tamu.edu	Solar radiation, relative humidity, wind speed	–
Google Earth image	https://www.earth.google.com	Road density	30 * 30 m
Hydrological data and Sedimentological data	https://cwc.gov.in/en, https://wbiwd.gov.in https://www.dvc.gov.in/	Reservoir details, discharge, sediment yield, water holding capacity	–

e-government websites. Key data sources include hydrological records from government agencies, remote sensing data from satellite imagery, digital elevation models (DEMs), land use/land cover (LULC) datasets, and sediment yield data from relevant studies. Additionally, geospatial data, including soil properties, catchment boundaries, and climate data, were obtained from government databases and research institutions. These diverse data sources are crucial for analyzing the spatial and temporal variations in sediment dynamics, soil erosion, sediment delivery ratio, and channel connectivity within the Damodar River Basin (DRB).

1.8 Existing Literature

Among the extensive literature on fluvial connectivity, soil erosion, sediment yield, erosional hotspots, reservoir sedimentation, and trapping efficiency, the following are particularly noteworthy. We have already discussed these concepts in a global context. Now, we will focus on the literature specific to the Damodar River Basin. The Damodar River Basin (DRB) is highly prone to water discharge and sediment supply, with a long history of recorded floods. These floods primarily occur due to the monsoon-dominated rainfall in the region. The Damodar Basin consists of monsoon-dominated rivers in India. Most geomorphic activities, including erosion, transportation, and deposition, occur during the monsoon season, primarily from July to September [45]. The Damodar Valley undergoes continuous adjustments in its aggradation and degradation processes due to regulated streamflow and the variability of monsoon rainfall [34]. As a result, most research studies in the Damodar River Basin focus on flood hydrology, flood morphology, paleohydrology, and hydrometeorology, emphasizing the basin's dynamic flood regime and climatic influences [4, 15, 17, 31–37, 45, 46, 52, 57, 58, 73, 74, 76, 77, 81] With the impacts of climate change, a rapidly growing population, land-use and land-cover changes, and increasing land pressure, the Damodar fluvial system has experienced significant disruptions. Reservoirs and dams have severely impaired channel connectivity throughout the system. Massive soil erosion and excessive sediment load in the upper Damodar catchments have led to critical issues, including reduced reservoir storage capacity and diminished sediment supply to the lower basin. In this context, precise estimation of inland sediment yield, soil erosion rates, discharge capacity, and connectivity is essential for effective management and planning. Various researchers have studied soil erosion and sediment yield in the Damodar River Basin [47–50, 70, 71, 78, 88], focusing on the impacts of land use changes, monsoon rainfall, and anthropogenic activities on sediment dynamics. The current work addresses these issues through mathematical modeling and the modern Remote Sensing technique.

1.9 Relevance of the Present Book

The relevance of this book lies in its contribution to understanding and addressing critical reservoir sedimentation issues within the Damodar River Basin (DRB), with broader implications for global watershed management. By focusing on the Damodar River Basin (DRB), this work addresses pressing challenges associated with sedimentation in reservoirs, soil erosion, and sediment connectivity. The investigations provided in this book provide valuable insights into identifying soil erosion hotspots, quantifying sediment yield, and understanding source-to-sink connectivity patterns, which are essential for effective watershed and dam management. This study bridges the intricate relationships between soil erosion (SE), sediment yield (SY), fluvial discharge (Q), soil erosion hotspots (SEH), reservoir sedimentation, and channel connectivity index (IC). By applying advanced mathematical modeling techniques, the research provides precise estimation and validation of sediment budgets, SE, SY, SEH, and sediment delivery ratios (SDR), ensuring scientific rigor and reliability.

A key focus of the research in this book is the identification of sub-watersheds or areas that contribute disproportionately to sediment inflow into major reservoirs such as Maithon, Panchet, and Tenughat. The analysis of fluvial channel connectivity highlights areas with high SDR, which are critical for understanding sediment transport pathways and their eventual deposition in reservoirs. This comprehensive approach not only quantifies sediment dynamics but also provides actionable insights for targeted sediment management interventions. This study integrates advanced geospatial techniques, hydrological modeling, and sediment delivery ratio (SDR) analysis to bridge the knowledge gap in sediment management strategies. By employing state-of-the-art methodologies such as SWAT and sediment connectivity indices, the research develops a holistic approach to assessing sediment transport and its implications for reservoir sustainability.

Ultimately, the investigation presented in this book provides a robust framework for addressing sedimentation challenges, contributing to the global discourse on watershed sustainability and river health restoration. It holds the potential to guide policymakers, resource managers, and researchers in implementing effective solutions to sedimentation problems, making it a meaningful step toward safeguarding river systems and their dependent ecosystems. This study not only advances theoretical understanding by refining concepts such as fluvial connectivity and soil erosion hotspots but also offers actionable recommendations for policymakers and water resource managers. The work is particularly relevant in the context of climate change, where intensified rainfall events and anthropogenic pressures exacerbate sedimentation and erosion challenges. Overall, this study underscores the urgent need for integrated sediment management frameworks to ensure the long-term functionality of dams and the sustainable management of river basins, making it a significant contribution to both regional and global water resource management initiatives.

1.10 Scope and Organization of This Book

This book is a comprehensive exploration of sediment dynamics, fluvial connectivity, and their implications for dam catchment management within the Damodar River Basin (DRB). It provides a multidisciplinary framework combining hydrological modeling, geospatial analysis, and field observations to examine sediment sources, transport pathways, and deposition patterns, emphasizing their relevance amid climate change and human interventions. With a focus on practical solutions, the book integrates advanced techniques like SWAT modeling, sediment delivery ratio (SDR) analysis, and connectivity indices (IC) to support watershed management and sustainable reservoir operations. The scope of this book extends beyond theoretical discussions, offering practical solutions for watershed management, sediment mitigation, and sustainable reservoir operations. It caters to a wide audience, including researchers, water resource managers, policymakers, and students, by presenting advanced techniques such as SWAT modeling, sediment delivery ratio analysis, and connectivity indices. Organized into nine chapters, it systematically covers the study objectives, physical attributes of the DRB, sediment and water dynamics, soil erosion hotspots, sediment connectivity, land use impacts, and actionable management strategies, concluding with key insights and future research directions. Designed for researchers, policymakers, and students, this book bridges theoretical understanding and practical applications, offering valuable contributions to sediment management and water resource sustainability.

1.11 Conclusion

This chapter offers a comprehensive analysis of sediment dynamics within the Damodar River Basin, focusing on the interconnected processes of fluvial connectivity, soil erosion hotspots, and reservoir sedimentation. By integrating global perspectives with regional insights, the conceptual framework effectively addresses sediment transport mechanisms from source areas to depositional sinks, emphasizing the influence of river regulation and human activities on natural sediment pathways. Through the examination of the Damodar River Basin, the research highlights the critical importance of identifying sediment sources and understanding their connectivity to large dams. This has significant implications for watershed management, soil conservation, and the sustainable operation of dams. Utilizing advanced methodologies and diverse datasets, the study quantifies sediment production, evaluates connectivity, and proposes practical solutions to mitigate sedimentation challenges. Ultimately, this work bridges theoretical concepts with practical applications, contributing valuable knowledge to sediment management strategies in regulated river basins.

References

1. Adamo N, Al-Ansari N, Sissakian V, Laue J, Knutsson S (2020) Dam safety: general considerations. J Earth Sci Geotech Eng 10(6):1–21. https://urn.kb.se/resolve?urn=urn:nbn:se:ltu:diva-79785
2. Admas BF, Gashaw T, Adem AA, Worqlul AW, Dile YT, Molla E (2022) Identification of soil erosion hot-spot areas for prioritization of conservation measures using the SWAT model in Ribb watershed, Ethiopia. Resour Environ Sustain 8(April):100059. https://doi.org/10.1016/j.resenv.2022.100059
3. Atristain M, von Schiller D, Larrañaga A, Elosegi A (2023) Short-term effects of a large dam decommissioning on biofilm structure and functioning. Restor Ecol 31(1). https://doi.org/10.1111/rec.13779
4. Bhattacharyya K (2011) The lower Damodar River, India. Springer, Netherlands. https://doi.org/10.1007/978-94-007-0467-1
5. Bhattacharyya K, Singh VP (2019) Reservoir sedimentation: assessment and environmental controls. CRC Press. https://www.routledge.com/Reservoir-Sedimentation-Assessment-and-Environmental-Controls/Bhattacharyya-Singh/p/book/9781138493636
6. Borrelli P, Alewell C, Alvarez P, Anache JAA, Baartman J, Ballabio C, Bezak N, Biddoccu M, Cerdà A, Chalise D, Chen S, Chen W, De Girolamo AM, Gessesse GD, Deumlich D, Diodato N, Efthimiou N, Erpul G, Fiener P, Panagos P et al (2021) Soil erosion modelling: a

global review and statistical analysis. Sci Total Environ 780:146494. https://doi.org/10.1016/j.scitotenv.2021.146494
7. Borselli L, Cassi P, Torri D (2008) Prolegomena to sediment and flow connectivity in the landscape: a GIS and field numerical assessment. CATENA 75(3):268–277. https://doi.org/10.1016/j.catena.2008.07.006
8. Bracken LJ, Croke J (2007) The concept of hydrological connectivity and its contribution to understanding runoff-dominated geomorphic systems. Hydrol Process 21(13):1749–1763. https://doi.org/10.1002/hyp.6313
9. Bracken LJ, Turnbull L, Wainwright J, Bogaart P (2015) Sediment connectivity: a framework for understanding sediment transfer at multiple scales. Earth Surf Proc Land 40(2):177–188. https://doi.org/10.1002/esp.3635
10. Bracken LJ, Wainwright J, Ali GA, Tetzlaff D, Smith MW, Reaney SM, Roy AG (2013) Concepts of hydrological connectivity: research approaches, pathways and future agendas. Earth Sci Rev 119:17–34. https://doi.org/10.1016/j.earscirev.2013.02.001
11. Brierley G, Fryirs K, Jain V (2006) Landscape connectivity: the geographic basis of geomorphic applications. Area 38(2):165–174. https://doi.org/10.1111/j.1475-4762.2006.00671.x
12. Calle M, Calle J, Alho P, Benito G (2020) Inferring sediment transfers and functional connectivity of rivers from repeat topographic surveys. Earth Surf Proc Land 45(3):681–693. https://doi.org/10.1002/esp.4765
13. Cavalli M, Trevisani S, Comiti F, Marchi L (2013) Geomorphometric assessment of spatial sediment connectivity in small Alpine catchments. Geomorphology 188:31–41. https://doi.org/10.1016/j.geomorph.2012.05.007
14. Cavalli M, Vericat D, Pereira P (2019) Mapping water and sediment connectivity. Sci Total Environ 673(April):763–767. https://doi.org/10.1016/j.scitotenv.2019.04.071
15. Chandra S (2003) INDIA: Flood management—Damodar river basin. World Meteorological Organization and Associated Programme on Flood Management, Integrated Flood Management-Case Study. www.apfm.info/pdf/case_studies/cs_india.pdf
16. Charlton R (2007) Fundamentals of fluvial geomorphology. In: Fundamentals of fluvial geomorphology. Routledge. https://doi.org/10.4324/9780203371084
17. Chatterjee SP (1967) Damodar valley planning atlas
18. Chaudhuri D (2017) Empirical approaches in prediction of reservoir sediment distribution—an experience of 57 reservoirs in the USA and India. Int J Sedim Res 32(2):260–276. https://doi.org/10.1016/j.ijsrc.2017.04.001
19. Chorley RJ, Schumm SA, Sugden DE (1985) Geomorphology. Methuen, London
20. Covino T (2017) Hydrologic connectivity as a framework for understanding biogeochemical flux through watersheds and along fluvial networks. Geomorphology 277(September 2016):133–144. https://doi.org/10.1016/j.geomorph.2016.09.030
21. Crema S, Cavalli M (2018) SedInConnect: a stand-alone, free and open source tool for the assessment of sediment connectivity. Comput Geosci 111:39–45. https://doi.org/10.1016/j.cageo.2017.10.009
22. Deng X, Xu Y, Han L, Song S, Xu G, Xiang J (2018) Spatial-temporal changes in the longitudinal functional connectivity of river systems in the Taihu Plain, China. J Hydrol 566(September):846–859. https://doi.org/10.1016/j.jhydrol.2018.09.060
23. Fang S, Chen M, Li J, Zhang L, Zhang Y, Yang C, Wang H (2024) Evaluation and optimization of urban hydrological connectivity based on graph theory: a case study in Chengdu, China. Ecol Inform 82(24):102749. https://doi.org/10.1016/j.ecoinf.2024.102749
24. Foteh R, Garg V, Nikam BR, Khadatare MY, Aggarwal SP, Kumar AS (2018) Reservoir sedimentation assessment through remote sensing and hydrological modelling. J Indian Soc Rem Sens 46(11):1893–1905. https://doi.org/10.1007/s12524-018-0843-6
25. Fryirs KA, Brierley GJ (2012) Geomorphic analysis of river systems. In: Geomorphic analysis of river systems. Wiley. https://doi.org/10.1002/9781118305454
26. Fryirs KA (2013) (Dis)Connectivity in catchment sediment cascades: a fresh look at the sediment delivery problem. Earth Surf Proc Land 38(1):30–46. https://doi.org/10.1002/esp.3242

27. Fryirs KA, Brierley GJ, Preston NJ, Spencer J (2007) Catchment-scale (dis)connectivity in sediment flux in the upper Hunter catchment, New South Wales, Australia. Geomorphology 84(3–4):297–316. https://doi.org/10.1016/j.geomorph.2006.01.044
28. Fu KD, He DM, Lu XX (2008) Sedimentation in the Manwan reservoir in the Upper Mekong and its downstream impacts. Quatern Int 186(1):91–99. https://doi.org/10.1016/j.quaint.2007.09.041
29. Getis A, Ord JK (1992) The analysis of spatial association by use of distance statistics. Geogr Anal 24(3):189–206. https://doi.org/10.1111/j.1538-4632.1992.tb00261.x
30. Getnet T, Mulu A (2021) Assessment of soil erosion rate and hotspot areas using RUSLE and multi-criteria evaluation technique at Jedeb watershed, Upper Blue Nile, Amhara Region, Ethiopia. Environ. Challenges 4(February):100174. https://doi.org/10.1016/j.envc.2021.100174
31. Ghosh D, Sheet S, Banerjee M, Karmakar M, Mandal M (2022) Flood characteristics and dynamics of sediment environment during Anthropocene: experience of the lower Damodar river, India. Sustain Water Resour Manage 8(2):58. https://doi.org/10.1007/s40899-022-00644-x
32. Ghosh S (2011) Hydrological changes and their impact on fluvial environment of the lower damodar basin over a period of fifty years of damming The Mighty Damodar River in Eastern India. Procedia Soc Behav Sci 19:511–519. https://doi.org/10.1016/j.sbspro.2011.05.163
33. Ghosh S, Guchhait SK (2014) Hydrogeomorphic variability due to dam constructions and emerging problems: a case study of Damodar River, West Bengal, India. Environ Dev Sustain 16(3):769–796. https://doi.org/10.1007/s10668-013-9494-5
34. Ghosh S, Guchhait SK (2014) Palaeoenvironmental significance of fluvial facies and archives of Late Quaternary deposits in the floodplain of Damodar River, India. Arab J Geosci 7(10):4145–4161. https://doi.org/10.1007/s12517-013-1079-y
35. Ghosh S, Guchhait SK (2016) Dam-induced changes in flood hydrology and flood frequency of Tropical river: a study in Damodar River of West Bengal, India. Arab J Geosci 9(2):90. https://doi.org/10.1007/s12517-015-2046-6
36. Ghosh S, Mistri B (2012) Investigating the causes of floods in Damodar River of India: a geographical perspective. Indian J Geomorphol 37–49
37. Ghosh S, Mistri B (2015) Geographic concerns on flood climate and flood hydrology in monsoon-dominated Damodar River basin, Eastern India. Geogr J 2015:1–16. https://doi.org/10.1155/2015/486740
38. Graf WL (2006) Downstream hydrologic and geomorphic effects of large dams on American rivers. Geomorphology 79(3–4):336–360. https://doi.org/10.1016/j.geomorph.2006.06.022
39. Guo L, Liu R, Men C, Wang Q, Miao Y, Shoaib M, Wang Y, Jiao L, Zhang Y (2021) Multiscale spatiotemporal characteristics of landscape patterns, hotspots, and influencing factors for soil erosion. Sci Total Environ 779:146474. https://doi.org/10.1016/j.scitotenv.2021.146474
40. Habel M, Mechkin K, Podgorska K, Saunes M, Babiński Z, Chalov S, Absalon D, Podgórski Z (2020) Dam and reservoir removal projects: a mix of social—ecological trends and cost—cutting attitudes. Sci Rep 1–16. https://doi.org/10.1038/s41598-020-76158-3
41. Heckmann T, Schwanghart W (2013) Geomorphic coupling and sediment connectivity in an alpine catchment—exploring sediment cascades using graph theory. Geomorphology 182:89–103. https://doi.org/10.1016/j.geomorph.2012.10.033
42. Hooke J (2003) Coarse sediment connectivity in river channel systems: a conceptual framework and methodology. Geomorphology 56(1–2):79–94. https://doi.org/10.1016/S0169-555X(03)00047-3
43. ICOLD (2023) World register of dams-general synthesis. ICOLD. https://www.icold-cigb.org/GB/world_register/database_presentation.asp
44. Jing Y, Zhao Q, Lu M, Wang A, Yu J, Liu Y, Ding S (2022) Effects of road and river networks on sediment connectivity in mountainous watersheds. Sci Total Environ 826:154189. https://doi.org/10.1016/j.scitotenv.2022.154189
45. Kale VS (2003) Geomorphic effects of monsoon floods on Indian Rivers. In: Flood problem and management in South Asia. Springer, Netherlands, pp 65–84. https://doi.org/10.1007/978-94-017-0137-2_3

46. Kale VS (2005) Fluvial hydrology and geomorphology of Monsoon-dominated Indian rivers. Revista Brasileira de Geomorfologia 6(1):63–73. https://doi.org/10.20502/rbg.v6i1.39
47. Karan SK, Ghosh S, Samadder SR (2019) Identification of spatially distributed hotspots for soil loss and erosion potential in mining areas of Upper Damodar Basin, India. CATENA 182(June):104144. https://doi.org/10.1016/j.catena.2019.104144
48. Kumar S, Mishra A, Raghuwanshi NS (2012). Estimating catchment sediment yield, reservoir sedimentation and reservoir effective life using SWAT Model. In: Proceedings of SWAT international conference, pp 18–20
49. Kumar S, Mishra A, Singh UK (2023) Assessment of land cover changes and climate variability effects on catchment hydrology using a physically distributed model. Sustainability (Switzerland) 15(13). https://doi.org/10.3390/su151310304
50. Kumar S, Raghuwanshi NS, Mishra A (2015) Identification and management of critical erosion watersheds for improving reservoir life using hydrological modeling. Sustain Water Resour Manage 1(1):57–70. https://doi.org/10.1007/s40899-015-0005-8
51. Kummu M, Varis O (2007) Sediment-related impacts due to upstream reservoir trapping , the lower Mekong River. 85:275–293. https://doi.org/10.1016/j.geomorph.2006.03.024
52. Lahiri-Dutt K (2006) State and the community in water management case of the Damodar Valley Corporation, India. In: International symposium on community activities for the conservation of water environment, pp 1–9
53. Lemma H, Frankl A, van Griensven A, Poesen J, Adgo E, Nyssen J (2019) Identifying erosion hotspots in Lake Tana Basin from a multisite soil and water assessment tool validation: opportunity for land managers. Land Degrad Dev 30(12):1449–1467. https://doi.org/10.1002/ldr.3332
54. Liu C, Chen Y, Huang W, Fang G, Li Z, Zhu C, Liu Y (2024) Climate warming positively affects hydrological connectivity of typical inland river in arid Central Asia. Npj Clim Atmos Sci 7(1). https://doi.org/10.1038/s41612-024-00800-4
55. Liu Y, Cui B, Du J, Wang Q, Yu S, Yang W (2020) A method for evaluating the longitudinal functional connectivity of a river–lake–marsh system and its application in China. Hydrol Process 34(26):5278–5297. https://doi.org/10.1002/hyp.13946
56. Maiti R (2023) Geotectonics and geomorphology an insight into process—form relationship. Nabodaya Publication
57. Majumdar A, Ghosh S, Dasgupta A, Seth D (2012) Analysis reservoir sedimentation of Panchet Dam, India using remote sensing and GIS. Panchakotessays, pp 82–95
58. Majumder M, Roy P, Mazumdar A (2010) An introduction and current trends of damodar and Rupnarayan river network. In: Impact of climate change on natural resource management. Springer Netherlands, pp 461–480. https://doi.org/10.1007/978-90-481-3581-3_25
59. Matanzima J, Mosuoe-tsietsi T, Matanzima J (2023) A complex balance: assessing perspectives on decommissioning large dams to restore river ecosystems. Water Int 48(5):615–630. https://doi.org/10.1080/02508060.2023.2234225
60. Mishra K, Sinha R, Jain V, Nepal S, Uddin K (2019) Towards the assessment of sediment connectivity in a large Himalayan river basin. Sci Total Environ 661:251–265. https://doi.org/10.1016/j.scitotenv.2019.01.118
61. Mu J, Wu Y, Qi P, Chen L, Yuan Y, Wu H (2024) Spatial and temporal change of hydrological connectivity in the Wusuli River Basin. J Hydrol Regional Stud 53(December 2023):101814. https://doi.org/10.1016/j.ejrh.2024.101814
62. Nilsson C, Ekblad A, Gardfjell M, Carlberg B (1991) Long-term effects of river regulation on river margin vegetation. J Appl Ecol 28(3):963–987. https://doi.org/10.2307/2404220
63. Niu Y, Shah FA (2021) Economics of optimal reservoir capacity determination, sediment management, and dam decommissioning. Water Resour Res 57(7):1–18. https://doi.org/10.1029/2020WR028198
64. Obialor CA, Okeke OC, Onunkwo AA, Fagorite VI, Ehujuo (2019) Reservoir sedimentation: causes, effects and mitigation. Int J Adv Acad Res Sci 5(10):2488–9849
65. Perera D, Williams S, Smakhtin V (2022) Present and future losses of storage in large reservoirs due to sedimentation: a country-wise global assessment. Sustainability 15(1):219. https://doi.org/10.3390/su15010219

66. Persichillo MG, Bordoni M, Cavalli M, Crema S, Meisina C (2018) The role of human activities on sediment connectivity of shallow landslides. Catena 160(August 2016):261–274. https://doi.org/10.1016/j.catena.2017.09.025
67. Peters DL, Prowse TD (2001) Regulation effects on the lower Peace River, Canada. Hydrol Proces 15(16):3181–3194. https://doi.org/10.1002/hyp.321
68. Poeppl RE, Keesstra SD, Maroulis J (2017) A conceptual connectivity framework for understanding geomorphic change in human-impacted fluvial systems. Geomorphology 277:237–250. https://doi.org/10.1016/j.geomorph.2016.07.033
69. Rahmani V, Kastens J, DeNoyelles F, Jakubauskas M, Martinko E, Huggins D, Gnau C, Liechti P, Campbell S, Callihan R, Blackwood A (2018) Examining storage capacity loss and sedimentation rate of large reservoirs in the central U.S. Great Plains. Water 10(2):190. https://doi.org/10.3390/w10020190
70. Rajbanshi J, Bhattacharya S (2020) Assessment of soil erosion, sediment yield and basin specific controlling factors using RUSLE-SDR and PLSR approach in Konar river basin, India. J Hydrol 587:124935. https://doi.org/10.1016/j.jhydrol.2020.124935
71. Rajbanshi J, Bhattacharya S (2022) Modelling the impact of climate change on soil erosion and sediment yield: a case study in a sub-tropical catchment, India. Model Earth Syst Environ 8(1):689–711. https://doi.org/10.1007/s40808-021-01117-4
72. Rincón G, Solana-Gutiérrez J, Alonso C, Saura S, García de Jalón D (2017) Longitudinal connectivity loss in a riverine network: accounting for the likelihood of upstream and downstream movement across dams. Aquat Sci 79(3):573–585. https://doi.org/10.1007/s00027-017-0518-3
73. Rudra K (2002) Floods in West Bengal, 2000—causes and consequences. In: Changing environmental scenario. ACB Publications, pp 326–347
74. Rudra K (2018) Rivers of the Ganga-Brahmaputra-Meghna Delta. Springer International Publishing. https://doi.org/10.1007/978-3-319-76544-0
75. Saco PM, Rodríguez JF, Moreno-de las Heras M, Keesstra S, Azadi S, Sandi S, Baartman J, Rodrigo-Comino J, Rossi MJ (2020) Using hydrological connectivity to detect transitions and degradation thresholds: applications to dryland systems. CATENA 186(March):104354. https://doi.org/10.1016/j.catena.2019.104354
76. Saha SK (1979) River-Basin planning in the Damodar Valley of India. Geogr Rev 69(3):273–287
77. Sanyal J, Carbonneau P, Densmore AL (2013) Hydraulic routing of extreme floods in a large ungauged river and the estimation of associated uncertainties: a case study of the Damodar River, India. Nat Hazards 66(2):1153–1177. https://doi.org/10.1007/s11069-012-0540-7
78. Sanyal J, Densmore AL, Carbonneau P (2014) Analysing the effect of land-use/cover changes at sub-catchment levels on downstream flood peaks: A semi-distributed modelling approach with sparse data. CATENA 118:28–40. https://doi.org/10.1016/j.catena.2014.01.015
79. Schopper N, Mergili M, Frigerio S, Cavalli M, Poeppl R (2019) Analysis of lateral sediment connectivity and its connection to debris flow intensity patterns at different return periods in the Fella River system in northeastern Italy. Sci Total Environ 658:1586–1600. https://doi.org/10.1016/j.scitotenv.2018.12.288
80. Schumm SA (1977) The fluvial system. Wiley, New York
81. Sen PK (1991) Flood hazards and river bank erosion in the Lower Damodar Basin. Indian geomorphology, pp 95–108
82. Shafroth PB, Perry LG, Helfield JM, Chenoweth J, Brown RL (2024) Vegetation responses to large dam removal on the Elwha River, Washington, USA. Front Ecol Evol 12(February):1–20. https://doi.org/10.3389/fevo.2024.1272921
83. Stevaux JC, Martins DP, Meurer M (2009) Changes in a large regulated tropical river: the Paraná River downstream from the Porto Primavera Dam, Brazil. Geomorphology 113(3–4):230–238. https://doi.org/10.1016/j.geomorph.2009.03.015
84. Stieglitz M, Shaman J, McNamara J, Engel V, Shanley J, Kling GW (2003) An approach to understanding hydrologic connectivity on the hillslope and the implications for nutrient transport. Global Biogeochem Cycles 17(4):1–15. https://doi.org/10.1029/2003GB002041

85. Surian N (1999) Channel changes due to river regulation: the case of the Piave River, Italy. Earth Surface Proces Landforms 24(12):1135–1151. https://doi.org/10.1002/(SICI)1096-9837(199911)24:12%3c1135::AID-ESP40%3e3.0.CO;2-F
86. Syvitski JPM, Vörösmarty CJ, Kettner AJ, Green P (2005) Impact of humans on the flux of terrestrial sediment to the global coastal ocean. Science 308(5720):376–380. https://doi.org/10.1126/science.1109454
87. Uunila L, Church M (2014) The regulation of peace river: a case study for river management. In: The regulation of peace river. Wiley, vol 9781118906, pp 115–140. https://doi.org/10.1002/9781118906170.ch6
88. Verma RK, Pandey A, Mishra SK, Singh VP (2023) A procedure for assessment of environmental flows incorporating inter- and intra-annual variability in dam-regulated watersheds. Water Resour Manage 37(8):3259–3297. https://doi.org/10.1007/s11269-023-03502-3
89. Walker KF (1985) A review of the ecological effects of river regulation in Australia. Hydrobiologia 125(1):111–129. https://doi.org/10.1007/BF00045929
90. Wang H, Wu X, Bi N, Li S, Yuan P, Wang A, Syvitski JPM, Saito Y, Yang Z, Liu S, Nittrouer J (2017) Impacts of the dam-orientated water-sediment regulation scheme on the lower reaches and delta of the Yellow River (Huanghe): a review. In: Global and planetary change, vol 157. https://doi.org/10.1016/j.gloplacha.2017.08.005
91. Wang J, Yin W, Wang LX, Li NX, Hao R, Xu JF, Xiao HB, Zhong ZM, Zhang YP, Shi ZH (2022) Unraveling the effects of hydrological connectivity and landscape characteristics on reservoir water quality. J Hydrol 613(2022):128410. https://doi.org/10.1016/j.jhydrol.2022.128410
92. Ward JV (1989) The four-dimensional nature of lotic ecosystems. J N Am Benthol Soc 8(1):2–8. https://doi.org/10.2307/1467397
93. Wen L, Rogers K, Ling J, Saintilan N (2011) The impacts of river regulation and water diversion on the hydrological drought characteristics in the Lower Murrumbidgee River, Australia. J Hydrol 405(3–4):382–391. https://doi.org/10.1016/j.jhydrol.2011.05.037
94. Wohl E (2017) Connectivity in rivers. Progress in physical geography: earth and environment 41(3):345–362. https://doi.org/10.1177/0309133317714972
95. Wohl E (2020) Rivers in the landscape. Wiley. https://doi.org/10.1002/9781119535409
96. Wohl E, Magilligan FJ, Rathburn SL (2017) Introduction to the special issue: connectivity in geomorphology. Geomorphology 277:1–5. https://doi.org/10.1016/j.geomorph.2016.11.005
97. Wohl E, Rathburn S, Chignell S, Garrett K, Laurel D, Livers B, Patton A, Records R, Richards M, Schook DM, Sutfin NA, Wegener P (2017) Mapping longitudinal stream connectivity in the North St. Vrain Creek watershed of Colorado. Geomorphology 277:171–181. https://doi.org/10.1016/j.geomorph.2016.05.004
98. Yu Y, Feng J, Liu H, Wu C, Zhang J, Wang Z, Liu C, Zhao J, Rodrigo-Comino J (2023) Linking hydrological connectivity to sustainable watershed management in the Loess Plateau of China. Curr Opin Environ Sci Health 35(July):100493. https://doi.org/10.1016/j.coesh.2023.100493
99. Zhang Q, Qin W, Ding L, Cao W, Xu H, Liu Q, Shi C (2023) Impact of land use/cover change (LUCC) on sediment connectivity in small watersheds based on a revised index algorithm. Geoderma 438(December 2022):116622. https://doi.org/10.1016/j.geoderma.2023.116622
100. Zhang Y, Huang C, Zhang W, Chen J, Wang L (2021) The concept, approach, and future research of hydrological connectivity and its assessment at multiscales. Environ Sci Pollut Res 28(38):52724–52743. https://doi.org/10.1007/s11356-021-16148-8
101. Zhao D, Jiao Y, He K, Xiong D, Zhang B (2024) Evaluating the effect of natural-artificial linear landscape elements on flow and sediment connectivity in a typical agricultural terraced catchment, China. CATENA 234(189):107561. https://doi.org/10.1016/j.catena.2023.107561

Chapter 2
Baseline Information of Damodar River Basin

Abstract The Damodar River Basin (DRB) is one of the most significant sub-watersheds in Ganga Basin due to its socio-economic and hydro-geomorphic significance. The basin spreads over Jharkhand and West Bengal, covering approximately 23,370 km^2. Historically, the river has been affected by seasonal flooding. Several major dams have been built to control floods and regulate water resources. This chapter conducts a detailed baseline investigation of the Damodar River Basin with special reference to hydrological, geological, and environmental features. The basin is primarily underlain by Precambrian metamorphic rocks, including schists, gneisses, and quartzites, with coal-rich Gondwana formations dominating its middle and lower reaches. The upper basin, located in the Chota Nagpur Plateau, features rugged, hilly terrain composed of hard crystalline rocks, while the lower reaches are characterized by sedimentary formations, particularly the coal-rich Gondwana formations that contribute to extensive mining activities. The slopes in the upper reaches range from 10° to as steep as 71.45°, with the steepest slopes concentrated in the Pareshnath Hill region and the upper catchment of the Damodar River. The insights gained from this analysis provide a key understanding of water resource management, flood control strategies, sustainable development, ensuring hydro-geomorphic sustainability, and supporting local economy and social needs.

Keywords Jharkhand · Damodar River Basin · Relative relief · Dissected and Degradational Uplands · Drainage system · Slope · Vegetation · Chota Nagpur plateau

2.1 Introduction

The Damodar River Basin (DRB), a sub-watershed of the Ganga River system, is characterized by highly rain-fed, shallow, wide, and flashy seasonal tropical flows that maintain a fluctuating regime [1]. This chapter will cover several hydrological

S. A. Alam and R. Maiti, *Sediment Source to Sink: Deciphering Sediment Connectivity to Large Dams in Damodar River Basin*, Earth and Environmental Sciences Library,
https://doi.org/10.1007/978-3-031-87072-9_2

and physical components that were examined to understand the major geomorphological aspects. The entire DRB has been divided into 23 sub-basins using the Arc-hydro tool in ArcGIS 10.3 software. The Damodar River basin covers Jharkhand and West Bengal, passing through the districts of Chatra, Latehar, Lohardaga, Ranchi, Hazaribagh, Koderma, Giridih, Bokaro, Dhanbad, and Jamtara in Jharkhand, and Purulia, Bankura, Burdwan, Hooghly, and Howrah in West Bengal. Almost all the districts are included in the present investigation. Sediment yield (SY), soil erosion (SE), sediment connectivity index (SCI), surface runoff, fluvial discharge, and sediment delivery ratio (SDR) are directly impacted by various factors, including soil types, geological formations, geomorphology, drainage density (DD), land use/land cover (LULC), rainfall, slope, vegetation, relative relief (RR), and human activities, etc. Thus, these hydrological, physical, and land resource aspects are addressed in this chapter.

2.2 Geography of Damodar River Basin (DRB)

The Damodar River Basin (DRB) is a sub-watershed of the Ganga fluvial system in India. In its upstream reaches, the Damodar River is locally known as "Deonad Nadi," a name given by the people of Jharkhand [2, 3, 8]. Throughout history, the river has been referred to by various names, as chronicled by ancient cartographers, geographers, and travelers. The Damodar River is known locally as "Damuda," which combines the words "Damu," meaning "sacred," and "Da," meaning "water" [8]. Additionally, it is referred to as "womb" or "Udar," symbolizing "full of fire," which alludes to the river flowing through a coal-rich region [1]. The Damodar River flows through some of the world's richest mineral belts, making the basin an epicenter for significant mining and industrial activity. As a result of these actions, the DRB is now highly susceptible to substantial soil erosion. The Damodar drainage basin has a tadpole-like general shape, with the head facing west [1].

The catchment area of the Damodar River Basin (DRB) extends between 22°13′ N and 24°30′ N latitude and 84°40′ E and 88°15′ E longitude. The river flows for 540 km, originating from the Khamarpath Hill region in the Chota Nagpur plateau [9]. The DRB includes the districts of Ranchi, Hazaribagh, Dhanbad, Bokaro, Latehar, and Lohardaga in the state of Jharkhand. The main tributaries of the Damodar River Basin are the Barakar, Bokaro, Jamunia, Konar, Behra, and Khadia rivers. The Barakar River originates from the Koderma plateau and passes through the districts of Koderma, Hazaribagh, Giridih, and Jamtara before reaching the Damodar near Dishergarh. The primary tributary of the Damodar River is the Barakar, which merges with the Damodar approximately 5.5 km downstream of the Panchet Dam in West Bengal. The two additional tributaries Nunia and Sali join the Damodar River just downstream of the Barakar confluence. The Damodar River was bifurcated into two branches before joining the Bhagirathi-Hooghly River near Jamalpur in the Bardhaman district. The eastern branch known as the Amta Channel merges with the Hooghly River at Gorchumbak near Uluberia, while the other branch, called the

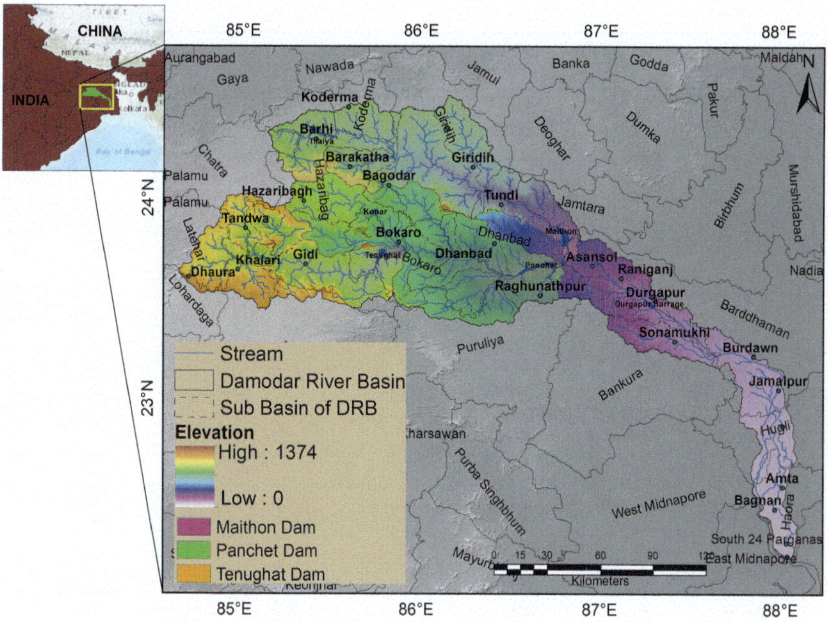

Fig. 2.1 Location map of the study area

Mundeswari, discharges into the Hooghly through the Rupnarayan Estuary [9]. The observation reveals that three-quarters of the total Damodar River Basin (DRB) area is situated in Jharkhand, while one-quarter is located in West Bengal (Fig. 2.1).

2.3 Geological Setup

Damodar River Basin (DRB) is composed of several geological formations. There are three main geological characteristics of this basin:

1. Peninsular shield mainly consisting of granites and gneisses of the Archaean Age
2. Gondwana tract filled up by sediment during the Permo-Carboniferous period (Older Alluvium)
3. Bengal basin was filled with newer Alluvium during the Tertiary and Quaternary periods.

The upper and middle catchment of the Damodar River Basin consists of the Chhotonagpur Gneissic Complex (Proterozoic), Upper Gondwana (Triassic to Early Jurassic), Lower Gondwana (Late Carboniferous to Permian), and unclassified metamorphic rocks (Archaean to Paleoproterozoic). The Damodar and Barakar rivers flow through the Chhotonagpur granite gneiss, as well as the Upper and Lower Gondwana series. In the lower stretch of the DRB, sandstone-Gondwana sedimentaries are found

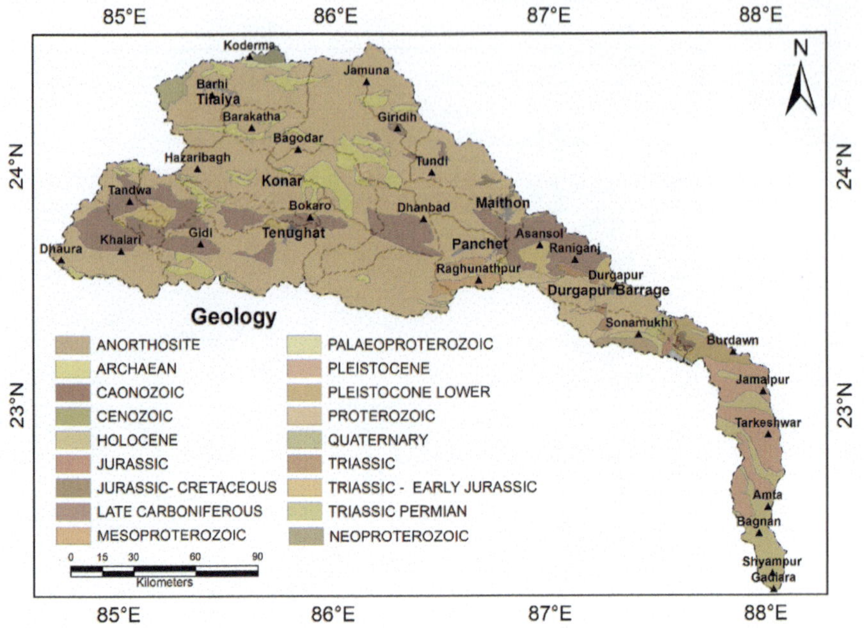

Fig. 2.2 Geological map of the Damodar River basin

near Asansol and Raniganj, along with Pleistocene and Holocene deposits extending from Jamalpur to the mouth of the Amta Channel (Fig. 2.2).

2.4 Physiographic Divisions

Physiographically, the Damodar River Basin (DRB) is divided into three separate microgeomorphic units. The basin's elevation range is classified into five categories: 0–124 m, 124–266 m, 266–386 m, 386–519 m, and 519–1374 m (Fig. 2.3). The altitudes of West Bengal range from 0 to 124 m above mean sea level (MSL), while the levels in Jharkhand range from 124 to 1374 m. The DRB's upper and middle catchment areas are distinguished by a variety of plateaus and uplands, including the Hazaribagh Plateau, Ranchi Plateau, Chhotanagpur Plateau, Koderma Plateau, Deoghar Upland, and Giridh Upland, with elevations varying from 200 to 1374 m above mean sea level. The three microgeomorphic units are as follows (Fig. 2.3):

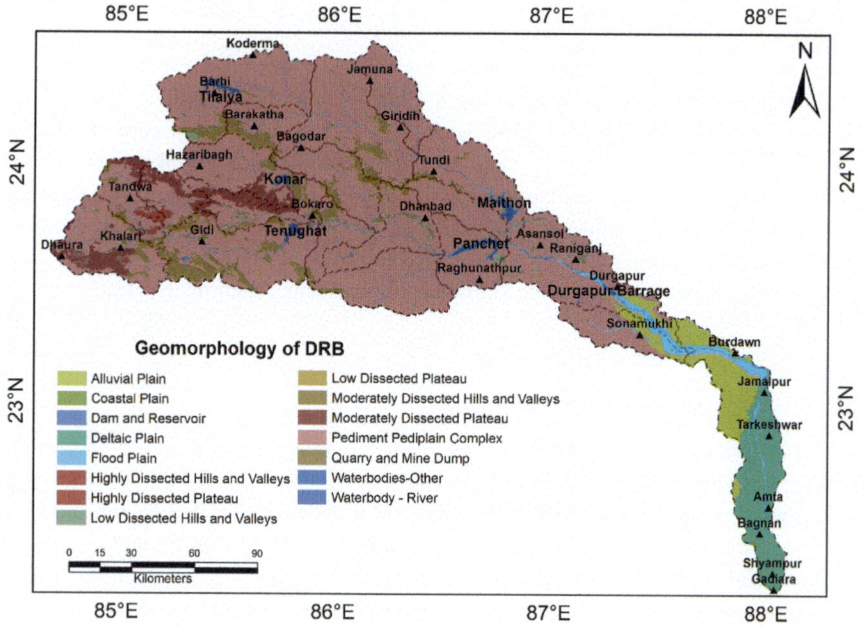

Fig. 2.3 A geomorphological map of the Damodar River Basin illustrates the distribution of various landforms within the basin

2.4.1 Plateau, Hills, and Ghats

The plateaus in the upper Damodar Basin are classified into three categories: upper, middle, and lower, respectively named as the Ranchi, Hazaribagh, and Koderma plateaus. The Ranchi Plateau, located in the upper basin, is characterized by high flat-lands interspersed with low rounded hills and monadnocks. The Hazaribagh Plateau has steep slopes and deeply incised valleys, resulting in a hard landscape with diverse plants and wildlife. Meanwhile, the Koderma Plateau is noted for its abundant mineral reserves and distinctive geological formations. The ghats along the river, notably those with steep banks, play an important role in managing sediment transport and affecting the flow patterns. These plateaus exhibit similar landform characteristics, featuring flat terrain, severely gullied areas, and isolated, low, rounded hills known as monadnocks, which are scattered throughout the landscape [7]. These landforms not only shape the physical environment but also impact local livelihoods, agriculture, and settlement patterns in the basin. Collectively, the plateaus, hills, and ghats contribute to the ecological and geomorphological complexity of the Damodar River Basin.

2.4.2 *Dissected and Degradational Uplands*

The dissected and degradational uplands extend across the middle section of the Damodar River Basin (DRB), encompassing areas in the Purulia, Bankura, Burdwan, and Birbhum districts of West Bengal, as well as the Ramgarh, Bokaro, Dhanbad, Giridh, Jamtara, and Dumka districts of Jharkhand. The distinctive landforms running from west to east are the dissected remnants of older peneplains characterized by steep slopes and significant relative relief (RR) around Pareshnath Hill. This hill area boasts the highest elevation in the Damodar River Basin, reaching 1374 m.

2.4.3 *Aggradational Flat Plains*

The aggradational flat plains of the Damodar River Basin are predominantly formed by sediment deposition from the Damodar and its tributaries. This region features low relief with elevations generally ranging from 10 to 20 m above mean sea level, characterized by fertile alluvial soils conducive to agriculture. This region is separated to the west from the dissected and degradational uplands by a line approximately joining Arambagh, Burdawn, and Bolpur. This region is characterized by monotonous flat alluvial plain land within 20 m of relative relief (RR). The lowest relative relief within 10 m lies near the south and southeast of Burdwan town, Shyamsundar, and the north side of Jamalpur. The plains are interspersed with various landforms, including natural levees and floodplains, shaped by seasonal flooding and sediment transport. These flat lands are crucial for supporting local livelihoods, primarily through agriculture and habitation.

2.5 Soil

The soil within the Damodar River Basin (DRB) has developed from the disintegration of rocks, with the lower catchment exhibiting soil formation through aggradational processes linked to the fluvial system. Minimal soil erosion suggests that this aggradational process significantly contributes to the soil's development. Various soil types are present in the basin (Fig. 2.4), including red soil, micaceous soil, sandy soil, black soil, alluvial soil, and laterite soil. According to the FAO classification, soils in the Damodar River Basin are categorized into groups such as Cambisols, Nitosols, Fluvisols, Luvisols, and Gleysols. In the lower catchment near Jamalpur, as well as the Amta and Mundeswari channels, the prevalent soil groups are Eutric Gleysols, Eutric Fluvisols, and Eutric Cambisols. Gleysols form under waterlogged conditions caused by rising groundwater and are typically used for rice cultivation. Cambisols lack a layer of accumulated clay, humus, soluble salts, or iron and aluminum oxides, often featuring medium-to-fine textured parent materials and high

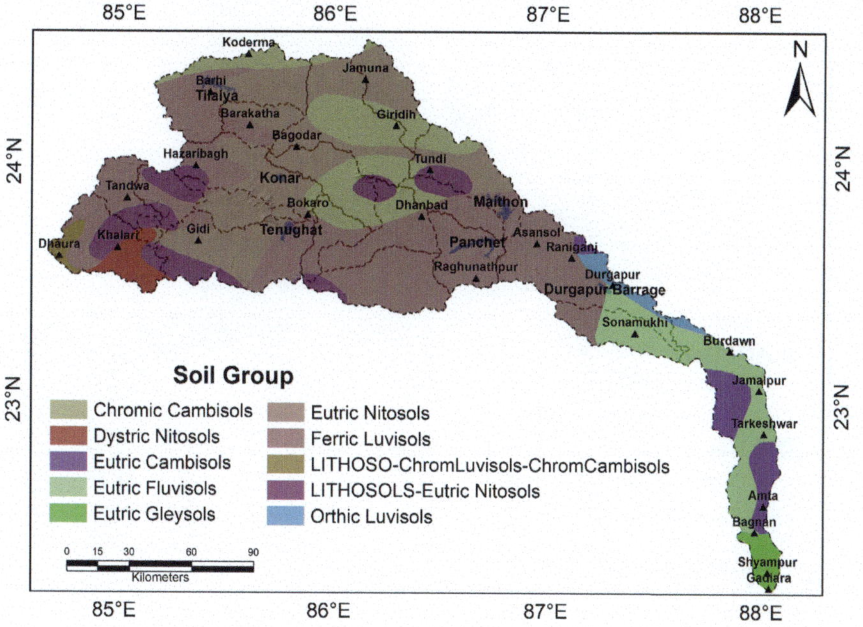

Fig. 2.4 Soil map of the Damodar River basin

erosion rates. Fluvisols are commonly found on level topographies that experience periodic flooding, occurring in floodplains, deltas, and coastal lowlands. In contrast, the upper and middle catchments of the DRB predominantly feature Eutric Nitosols, Ferric Luvisols, and Orthic Luvisols. Nitosols, found at higher altitudes, are particularly fertile due to their high nutrient content and deep, permeable structures, while Luvisols are nutrient-rich soils that develop in flat or gently sloping landscapes with good drainage conditions.

2.6 Drainage System

The Damodar River Basin (DRB) is situated within the Chotanagpur plateau and the Bengal deltaic plains, forming the northeastern section of India's peninsular plateau. The upper and middle catchments flow eastward and then southeastward as they enter the deltaic plains of Bengal below Raniganj. The Damodar River continues its course until Burdwan, where it abruptly shifts from an easterly to a southerly direction, ultimately joining the Bhagirathi-Hugli River at Gorchumbak near Uluberia (Fig. 2.5). The main tributary of the Damodar River is the Barakar, which originates near Padma in Hazaribagh district and flows approximately 225 kilom across the northern part of the Chotanagpur plateau in an easterly direction before bending slightly southeast to merge with the Damodar near Dishergarh in Asansol. Numerous tributaries and

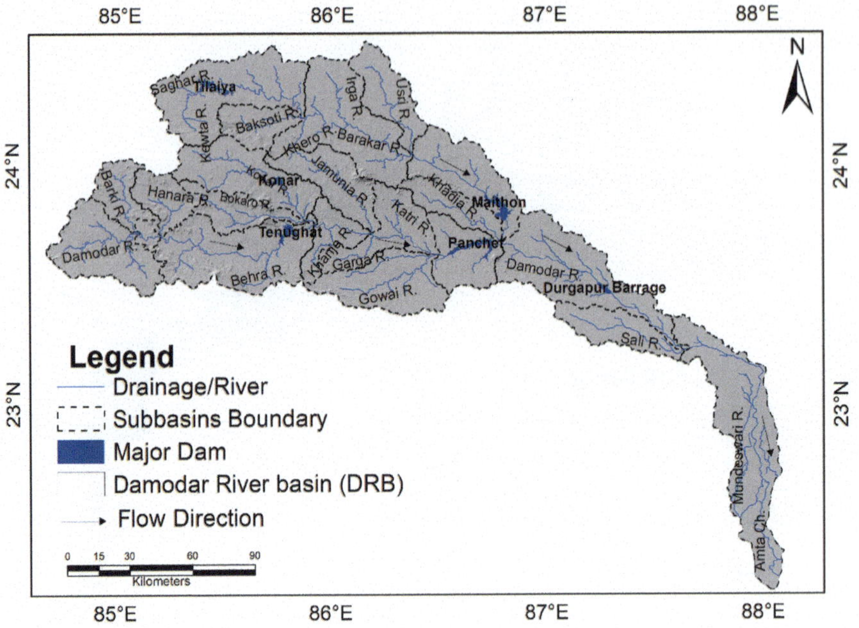

Fig. 2.5 A detailed drainage system of the Damodar River Basin includes the main Damodar River and its network of tributaries

sub-tributaries of the DRB, including Konar, Behra, Bokaro, Jamunia, Katri, Hahara, Khadia, Garga, Usri, Irga, and Ketwa, surround the Barakar. In the lower catchment of the Damodar basin, additional tributaries such as Nunia and Sali contribute to the river system.

2.7 Reservoir and Dams

The Damodar River Basin (DRB) has a long history of significant flood vulnerability, experiencing varying magnitudes of flooding approximately every 7–10 years. In response to this persistent challenge, the government initiated the construction of a series of dams modeled after the development strategies employed in the Tennessee Valley, USA. Consequently, the Damodar Valley Corporation (DVC) was established as a river basin organization (RBO) on July 7, 1948 [4]. The DVC oversees a network of five key dams [6]: the Tilaiya Dam (constructed in 1953) and the Maithon Dam (1957), both located on the Barakar River; the Konar Dam (1955) on the Konar River; the Panchet Dam (1959) on the Damodar River; and the Tenughat Dam (1978), also situated on the Damodar River [5]. Additionally, the Durgapur Barrage (1955) was constructed, and primarily managed through two reservoirs, Mithon and Panchet, to enhance flood control and water resource management within the basin.

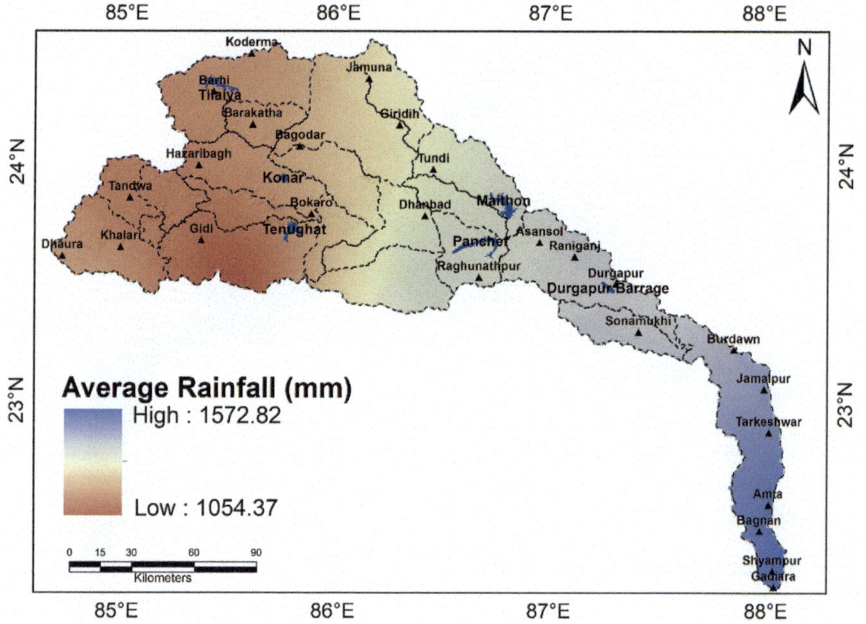

Fig. 2.6 An average rainfall distribution map of the Damodar catchment

2.8 Rainfall

Rainfall data for the Damodar River catchment area has been sourced from the India Meteorological Department (IMD) for the period of 40 years, from 1980 to 2020. The data from 28 stations were grouped and arithmetic averages were calculated to determine the mean rainfall across the basin (Fig. 2.6). Over the last four decades, the Damodar River Basin (DRB) has received average annual rainfall ranging from 1054 to 1572 mm, surpassing the national average of 1000 mm. The majority of this rainfall occurs between June and September. The upper and middle catchments receive an average of 1100–1300 mm, while the lower DRB experiences higher rainfall, ranging from 1300 to 1570 mm. The Barakar River basin, a key tributary, typically receives more rainfall than the Damodar River basin itself.

2.9 Relative Relief (RR)

The Damodar River Basin (DRB)'s relative relief (RR) was calculated utilizing a digital elevation model (DEM). Relative relief refers to the difference between the highest and lowest elevations within a certain area or grid, which is essential for comprehending the topography of a region. The RR throughout the DRB ranges from 0 to 601 m (Fig. 2.7). The central and southwest parts of the basin have the

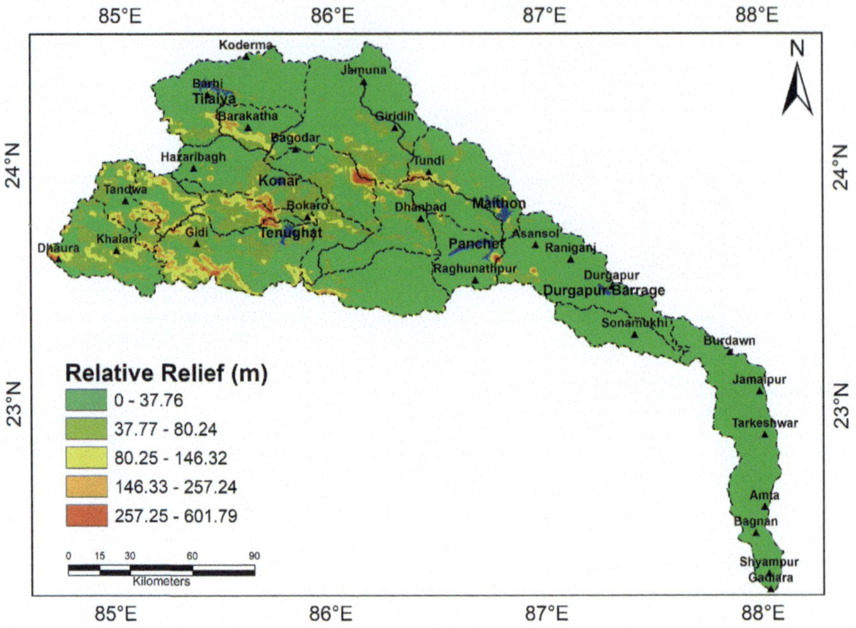

Fig. 2.7 Relative Relief (RR) map of the Damodar River Basin

highest RR values, whilst the lower catchment has the most lowered relief. The lowest relative relief is found in the flood plain of West Bengal, especially in the areas south of Burdwan and opposite the Jamalpur region. Relative relief ranging from 10 to 20 m is found below Raniganj, extending towards the mouths of the Amta and Mundeswari channels. The areas with RR between 20 and 40 m are located in the lower sections of the Barakar River Basin and the lower Damodar River Basin. The highest relative relief, between 40 and 601 m, is concentrated around Pareshnath Hill and the upper reaches of the Damodar basin.

2.10 Slope

Slope measures the rate of elevation change and plays a vital role in analyzing surface hydrology, geomorphological processes, and physical factors. Within the Damodar River Basin (DRB), slopes range from 0 to 71.45° (Fig. 2.8). More than 80% of the basin consists of slopes less than 10°, while the remaining 20% features steeper slopes between 10 and 71.45°. The steepest slopes, greater than 10°, are concentrated in the Pareshnath Hill region and the upper catchment of the Damodar River.

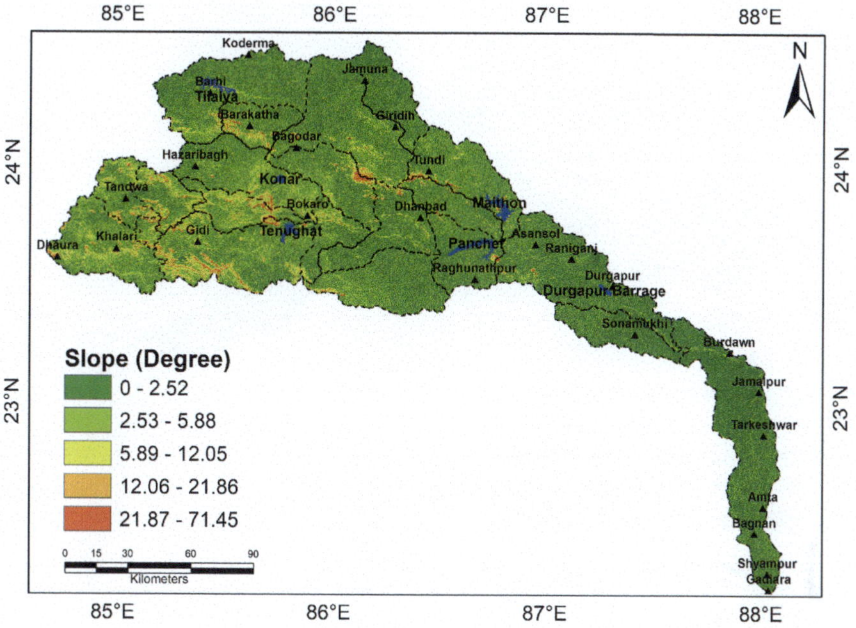

Fig. 2.8 Slope map of Damodar River Basin

2.11 Vegetation

The vegetation within the Damodar River Basin (DRB) has been assessed using the Normalized Difference Vegetation Index (NDVI). NDVI readings in the research area range from − 0.29 to 0.51 and were classified into five categories using the natural break classification method (Fig. 2.9). These classifications reflect various land cover types: Water Bodies (− 0.29 to 0.01), Bare Soil (0.01–0.13), Sparse Vegetation (0.13–0.18), Intermediate Green Vegetation (0.18–0.21), and Dense Vegetation (0.21–0.51). The middle and lower catchments of the Damodar Basin, including areas like Maithon Dam, Asansol, Raniganj, and Durgapur, exhibit significant bare soil. Dense vegetation concentrates in the upper catchment of the Damodar and Barakar basins, while intermediate green vegetation dominates the agricultural regions of the lower catchment.

2.12 Sub-Basins of Damodar River Basin

The Damodar River Basin (DRB) serves as a significant sub-basin of the Ganga River system, which demands a detailed understanding of its hydrological processes not just at the basin level but also at the sub-basin scale. Using SRTM DEM in ArcGIS, sub-basins were delineated to capture the spatial intricacies of the watershed. As per

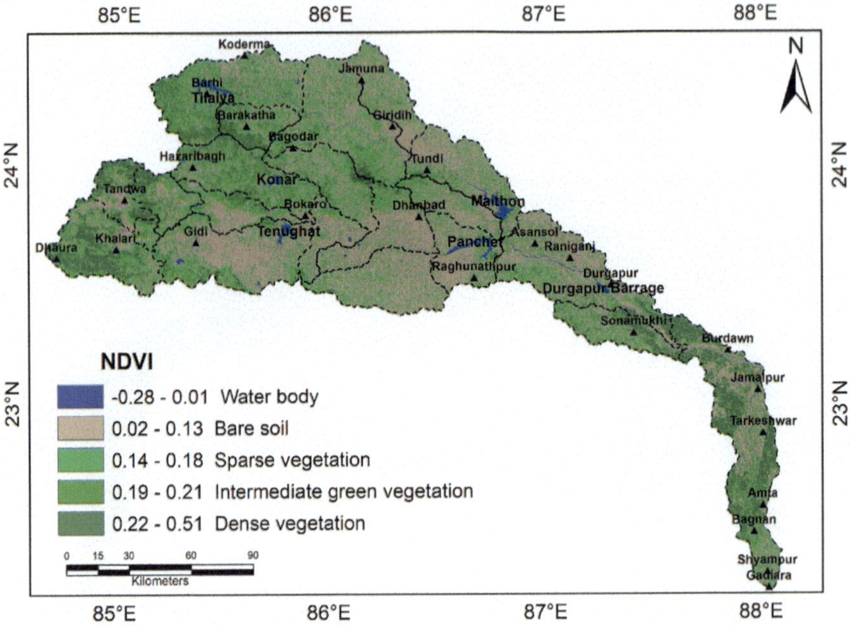

Fig. 2.9 The vegetation status of the Damodar River Basin refers to the distribution and types of vegetation found across the region

the Central Ground Water Board (CGWB), the DRB comprises approximately 64 sub-basins, with boundaries defined based on drainage patterns and catchment areas. Through ArcGIS 10.3 and the Arc-hydro tool, 23 sub-basins were identified: 5 within the Barakar River basin, 13 within the Damodar River basin extending to Panchet Dam, and 3 in the lower catchment region (Fig. 2.10). This hierarchical sub-division enhances the granularity of hydrological analysis across the entire basin.

2.13 Major Findings

The key findings of this chapter include:

- The Damodar River Basin (DRB) has been classified into 23 sub-basins based on its drainage network and elevation profiles. Sub-basins 1–20 are situated within the upper and middle catchment areas of the DRB, while sub-basins 21 to 23 are located in the lower catchment region.
- The elevational range of the DRB spans from 0 to 1374 m. Specifically, the upper and middle catchment areas, located in Jharkhand, extend from 124 to 1374 m, while the lower catchment area in West Bengal ranges from 0 to 124 m.

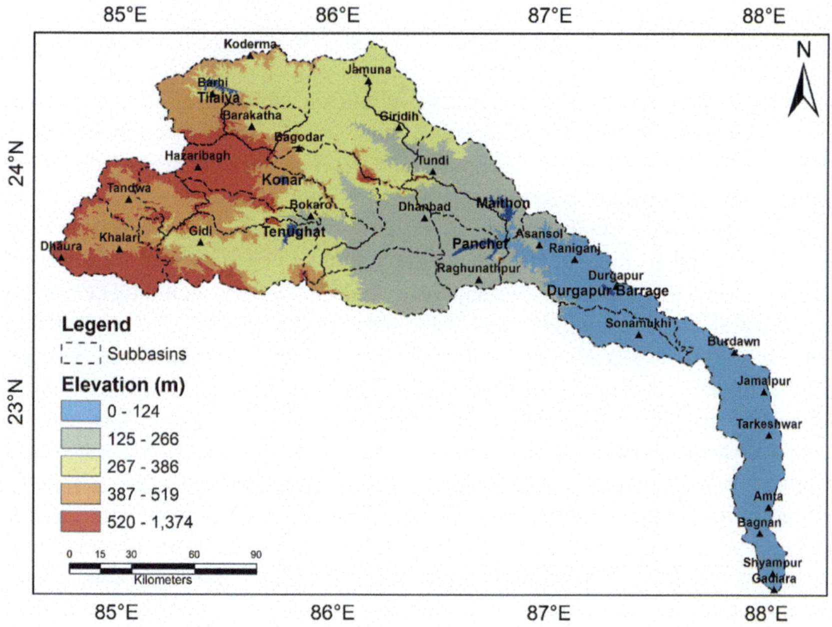

Fig. 2.10 The major sub-basins of the Damodar River Basin

- Geologically, the upper and middle catchments of the Damodar River Basin are characterized by rock formations including Archaean granitic gneisses, sandstones, shales, and Gondwana coal deposits. In contrast, the lower part of the basin is composed of the geology of the Bengal basin, which has been predominantly filled with newer alluvial sediments deposited during the Tertiary and Quaternary periods.
- The Damodar River Basin is characterized by four primary soil types. Red loam soils are prevalent in the upper and middle regions, while lateritic soils are found in the middle section. The upper catchment is distinguished by dark grey calcareous soils, whereas riverine alluvial soils dominate the lower catchment area [8].
- The estimated average annual rainfall over the Damodar River basin (DRB) from 1980 to 2020 varied from 1054 to 1572 mm.
- The Barakar River Basin experiences significantly higher rainfall than the Damodar River Basin, with the greatest precipitation recorded in the lower regions of the Barakar Basin.
- The Damodar Basin exhibits a relative relief (RR) ranging from 0 to 601 m, with the highest elevations found in the vicinity of Pareshnath Hill and the upper reaches of the basin, where relative relief values span from 40 to 601 m.
- The Damodar River Basin (DRB) features a slope gradient ranging from 0° to 71.45°. Notably, over 80% of the basin's area falls within a slope of less than 10°, while the remaining 20% exhibits slopes ranging from 10° to 71.45°.

2.14 Conclusion

The Damodar River Basin (DRB) serves as a significant sub-basin of the Ganga River system, comprises three micro-geomorphic units, and is characterized by a distinct distribution of soils: red and lateritic soils in the upper regions, along with dark grey soils, while the lower basin predominantly features riverine alluvial soils. This basin has been significantly impacted by various developmental activities, which have adversely affected its hydrological and morphological characteristics. The fluvial dynamics of the DRB have transformed due to multiple factors, including deforestation, industrial activities, mining, and dam construction. Notably, the construction of dams has dramatically altered the fluvial system of the Damodar River, with the upper and middle catchments being particularly vulnerable to soil erosion. Twenty-three sub-basins within the DRB exhibit varying degrees of impact due to their diverse hydrological, physical, human, and economic attributes. Certain sub-basins are significant contributors of sediment and soil erosion to the main channel, influencing reservoir sedimentation and dynamics in the lower catchment. This chapter considers a wide range of factors, including elevation, land use, rainfall, slope, soil composition, geology, vegetation, and physiographic characteristics. Thus, a sub-basin-level analysis is essential for a comprehensive understanding of the basin's dynamics. In conclusion, the interplay of physical, hydrological, physiographic, and anthropogenic factors is crucial for elucidating the processes of soil erosion (SE), sediment yield (SY), channel connectivity, fluvial discharge, and sediment delivery within the DRB across various spatial and temporal scales.

References

1. Bhattacharyya K (2011) The lower Damodar River, India. Springer, Netherlands, Dordrecht
2. Bhattacharyya K (1998) Applied geomorphological study in a controlled tropical river: the case of the Damodar between Panchet reservoir and Falta. PhD dissertation, The University of Burdwan, West Bengal, India
3. Bhattacharyya K (2000) Proper planning can save Damodar in Bengali (Nikut parikalpana Damodarke banchate pare). Bartaman, 17 Oct 2000
4. Chandra S (2003) India: flood management—Damodar River Basin. Retrieved from www.apfm.info/pdf/case_studies/cs_india.pdf. Accessed 13 Oct 2011 at 09:35 am
5. CWC (Central Water Commission) (2001) Compendium on silting of reservoirs in India. Government of India, New Delhi, India: Reservoir Sedimentation Directorate, CWC
6. DVC (2021) Reservoir overview & salient features. Damodar Val. Corp. https://www.dvc.gov.in/cms-web/details-pages/113
7. Mahadevan TM (2002) Geology of Bihar & Jharkhand. Geological Society of India, Bangalore, p 563
8. Mondal GC, Singh AK, Singh TB (2018) Damodar River basin: storehouse of Indian Coal. In: Singh DS (ed) The Indian Rivers. Springer Hydrogeology. https://doi.org/10.1007/978-981-10-2984-4_21
9. Rudra K (2018) Rivers of the Ganga-Brahmaputra-Meghna Delta. Springer International Publishing, Cham

Chapter 3
Estimating Sediment Yield (SY), Water Discharge and Surface Runoff: SWAT Model

Abstract Accurate estimation of sediment yield (SY), water discharge, and surface runoff is essential for effective water resource management in dam catchments. This chapter estimates Sediment Yield (SY), water discharge, and surface runoff in the Damodar River Basin, specifically focusing on the catchments of Maithon, Panchet, and Tenughat Dams, using the Soil and Water Assessment Tool (SWAT) model. The SWAT model simulates hydrological processes, including sediment transport, runoff generation, and discharge, based on environmental factors such as land use, soil characteristics, topography, and precipitation. Utilizing 30 years of hydro-climatic data, the model was calibrated and validated to simulate sediment yield, discharge, and runoff across various sub-watersheds. The model's performance was assessed using statistical measures such as R^2, RMSE, and MAE to evaluate the accuracy of simulated values compared to observed data. The model results revealed spatial and temporal variations in sediment transport and runoff dynamics, providing valuable insights into erosion-prone areas and sediment deposition zones within the basin. This chapter highlights the effectiveness of the SWAT model in predicting sediment dynamics and hydrological variables, providing critical information for discharge, soil erosion, and sediment yield rate in the Damodar River Basin. This chapter provides crucial information for enhancing the sustainability of water resources, improving sediment management strategies, and addressing sedimentation issues in the Damodar River Basin.

Keywords SWAT model · Surface runoff · Discharge · Sediment yield · Sub-basins · Land use land cover

3.1 Introduction

A watershed is a comprehensive system sustained by the interconnected processes of input, processing, and output. The precipitation acts as an input in this fluvial system where various activities of water working on the watershed act as a process and the fluvial discharge and sediment act as an output on the fluvial system. Water flowing

S. A. Alam and R. Maiti, *Sediment Source to Sink: Deciphering Sediment Connectivity to Large Dams in Damodar River Basin*, Earth and Environmental Sciences Library,
https://doi.org/10.1007/978-3-031-87072-9_3

through a fluvial system carries out three kinds of work: Erosion, transportation, and deposition. So, erosion, transportation, and deposition are associated with the discharge, sediment yield, and surface runoff. Thus, understanding the nature of discharge, sediment yield and surface runoff in the watershed is crucial for sustainable management planning and development. Inputs of sediment to the fluvial channel networks are highly variable over both space and time [2]. Water flowing through a channel on the surface is called runoff where the volume of water flowing through it at a certain point at a given time second is called discharge, whereas, sediment yield is the result of the effectiveness of surface runoff and fluvial discharge. Again, the competence and capacity of the channel are determined by the nature and volume of sediment in the channel.

Human modification and encroachments into nature affect the watershed system as it does in other natural systems. The modification or alteration has instantaneously affected the fluvial discharge regimes and surface runoff which, in turn, has directly affected the sediment yield of the system. Construction of dams, modification of channels, and transfer of water through the other channels result in serious challenges in river regulation, aquatic environment, fluvial environment, and channel connectivity. So, this chapter attempts to address the variability of fluvial discharge, sediment yield (SY), and surface runoff by using the ArcGIS interface soil water assessment tool (SWAT).

3.2 Conceptual Background of the SWAT Modelling

Over the past several decades, a considerable number of models were developed for soil erosion, fluvial discharge, and surface runoff. Simultaneously, the models involve more complexity to address real situations. With all these intricate things in mind, we try to use conceptual models to understand the diversity and complexity of sediment yield, discharge, and surface runoff. Conceptual models are the amalgamation of empirical and physical-based models [1].

The Soil and Water Assessment Tool (SWAT) is a conceptual model [1, 5] extensively used over the world for the resolution of the watershed-scale hydrological complexity and diversity. The Soil Water Assessment Tool (SWAT) is a comprehensive, Empirio-Physical, Continuous, Semi-Distributed, and Process-based integrated hydrological model that has been designed for the assessment of the influence of land management practice on hydrology, water quality, and sediment yield in the large, complex watersheds [1]. The SWAT model was developed by the US Department of Agricultural–Agricultural Research Service (USDA–ARS) at the Blackland Research and extension center in Temple, Texas, USA. As an integration of various hydrological models such as CREAMS (Chemical, Runoff, and Erosion from Agricultural Management Systems), GLEAMS (Groundwater Loading Effects on Agricultural Management Systems), EPIC (Erosion—Productivity Impact Calculator), SWRRB (Simulator for Water Research Basins) and ROTO (Routing Outputs and

Outlet). Finally, the SWRRB and ROTO models were merged into a single model called SWAT [1]

Since SWAT was built in the early 1990s, it has passed through continuous revision and development to expand its capabilities. This SWAT hydrological semi-distributed model has been extensively used for the simulation of time continuous sediment yield (SY), discharge (Qs), and Surface runoff in the ungagged large river basin. The various Windows interfaces have been developed over time. ArcSWAT interface, QSWAT interface [3], SWATplusR interface [9], etc.

3.3 SWAT Modelling Technique

The SWAT (Soil Water Assessment Tool) model is a hydrological processed-based model that assesses the water balance, surface runoff, climate change impact on land management, nutrient loss, sediment yield, and plant growth components. The model readily uses spatial and temporal data for simulating the water flows in groundwater and soil, nutrient cycling, sediment yield, pesticides, erosion, bacteria, and environmental influence on climate change [1, 12]. The SWAT model established the impact of climate change or land management practices on sediment yield, water discharge, and surface runoff. For the illustration of spatial heterogeneity of the watersheds, initially, the basin is classified into sub-basins, and then each sub-basin is subdivided into hydrological response units (HRUs) based on identical combinations of land use characteristics, soil types, and slope. The watershed hydrological cycle for SWAT simulation can be divided into two major phases: land phases and routing phases. The *land phases* determine the water, sediment, nutrients, and pesticides for every HRU within each sub-basin, and then loadings are calculated and aggregated to the main channel of the sub-basins. The *routing phases* control the water and sediment, etc. on the main channels and outlets [7]. In the SWAT model, hydrological calculations are based on the water balance equation in the soil profile, which includes precipitation, surface runoff, infiltration, percolation, groundwater flow, and evapotranspiration [1].

$$SW_t = SW_0 + \sum_{i=1}^{t}(R_{day} - Q_{surf} - E_a - W_{seep} - Q_{gw}) \tag{3.1}$$

where SW_t is the final soil water content at period t (mm), SW_0 is the initial soil water content (mm), *t* is the time (number of days), R_{day} is the amount of rainfall on *the ith* day (mm), Q_{surf} is the amount of surface runoff on the *ith* day (mm), E_a is the amount of evapotranspiration on the *ith* day (mm), W_{seep} is the amount of water giving recharge to groundwater from the soil profile on *the ith* day (mm), and Q_{gw} is the amount of return flow on the *ith* day (mm).

The SWAT model also simulates the sediment yield (SY) prediction on daily/monthly/yearly time step duration based on the MUSLE equation [10]. The MUSLE model is the modified version of the USLE model which was developed by [11]. The USLE model predicts the average annual gross soil erosion rate in particular areas as a function of rainfall energy. For the advancement of gross soil erosion to sediment yield predictions, the runoff factor was replaced by the rainfall energy factor in the MUSLE model. The significant factor of sediment yield is dependent on the transport capacity of runoff in particular areas [6]. The current version of the SWAT model allows the MUSLE model for the amount of sediment as a function of peak channel velocity and estimate sediment yield (SY) for every HRU in the watershed [10]. The modified universal soil loss equation (MUSLE) model follows the Eq. (3.2).

$$sed = 11.8 \cdot \left(Q_{surf} \cdot Q_{peak} \cdot area_{hru}\right)^{0.56} \cdot K_{USLE} \cdot C_{USLE} \cdot P_{USLE} \cdot LS_{USLE} \cdot CFRG \tag{3.2}$$

where sed is sediment yield (t/day), Q_{surf} is the daily runoff volume (mm H_2O/ha), Q_{peak} is the runoff peak discharge (m^3/s), $area$ is the area of the HRU (ha), K_{USLE} is the USLE soil erodibility factor (0.013 $\left(\text{tha}^{-1}\text{h}^{-1}\text{MJ}^{-1}\text{mm}^{-1}\right)$, C_{USLE} is the USLE cover and management factor, P_{USLE} is the USLE support practice factor, L_{USLE} is the USLE topographic factor and CFRG is the coarse fragment factor.

3.3.1 SWAT Model Input

The SWAT model was implemented in ArcGIS "Version 10.3" and an extension tool has been added as "ArcSWAT2012" in ArcGIS which facilitates the exercise of prepared data. This model requires intensive and precise hydrometeorological data for the simulation of various features in the watershed. The principal hydrometeorological data include meteorological data (Temperature, precipitation, wind speed, relative humidity, solar radiation), LULC (land use and land cover), and soil texture. The following data were selected for the modeling of SWAT input on the study of the Damodar River Basin (DRB).

(A) **DEM**: The digital elevation model (DEM) has been used for the SWAT modeling to extract the various topographic attributes, stream networks, and sub-catchment up to the HRUs level in this catchment. In this chapter, the SRTM (Shuttle Rader Thematic Mapping) 30-m DEM has been used which was downloaded from the NASA Earth data (https://earthdata.nasa.gov) website and used as input data in the modeling.

(B) **Soil map**: Fluvial discharge, surface runoff, infiltration, and soil erosion are significantly correlated with the soil textural properties and physio-chemical properties of an individual soil group. In this chapter, all properties of soil are coupled with one database for the simulation. In this study, the digital soil database (DSMW) was obtained from the Food and Agricultural Organisation of the United Nations (FAO–UN) (https://www.fao.org). All the soil attributes were added manually in the ArcSWAT database table.

(C) **Land Use Land Cover**: Land use land cover of this Damodar River basin (DRB) is obtained from this website (https://daac.ornl.gov/cgi-bin/dsviewer.pl?ds_id=1336). The overall classification accuracy is within an acceptable range of 94.46% and the kappa accuracy of 0.9445 for this LULC database [8]. This catchment region is divided into eleven land use and cover classifications. The most dominant land uses in this research region are forest and agricultural land. This LULC map serves as input data for the ArcSWAT model.

(D) **Climate**: The meteorological input data in ArcSWAT including minimum temperature, maximum temperature, solar radiation, relative humidity, wind speed, and precipitation are obtained from global weather data for SWAT (https://globalweather.tamu.edu). The SWAT weather generator tool also helps us to generate weather input and fill in missing values for simulation time.

3.3.2 SWAT Model Setup

ArcSWAT is one of the beneficial graphical user interface methods of SWAT which is applied to the Damodar River Basin (DRB) for discharge and sediment yield simulation. This DRB is divided into 23 sub-basins based on elevation range and their catchment areas. The ArcSWAT model is processed by various stages (Fig. 3.1). All input data are projected with a similar projection system (WGS 1984) for creating a watershed from DEM and classified into various sub-basins. The sub-basins were subdivided into Hydrological response units (HRUs). The HRU analysis tool in ArcSWAT supports loading the soil, land uses land cover, and slope with application to the threshold values of 10, 10, and 10% respectively. The delineated watershed and various land use land cover, soil, and slope layers are overlayed by 100% in the ArcSWAT tool in ArcGIS. In the second step, all the necessary climatic variables are fed to the weather generator tool in ArcSWAT interface by text (tab-delimited) format. Finally, all databases are linked to all data layers and processed by built-in mathematical formulas to simulate the various features like fluvial discharge, sediment yield surface runoff, etc. for every HRU. After all the above processes are completed, the total simulation period (including a 3-year warm-up period) is set to run from 1979 to 2013 (i.e. 34 years).

3.4 Water and Sediment Discharge: Damodar River Basin (DRB) as a Whole

The SWAT model has been applied to the Damodar River Basin (DRB) to simulate the discharge (Q), surface runoff, and sediment yield (SY) on a yearly and monthly basis. The catchment area of the Damodar River basin is apportioned with various landforms like hills, plateaus, and plain land. The greatest section of this catchment is covered by the plateau areas and is enormously affected by mining activities which

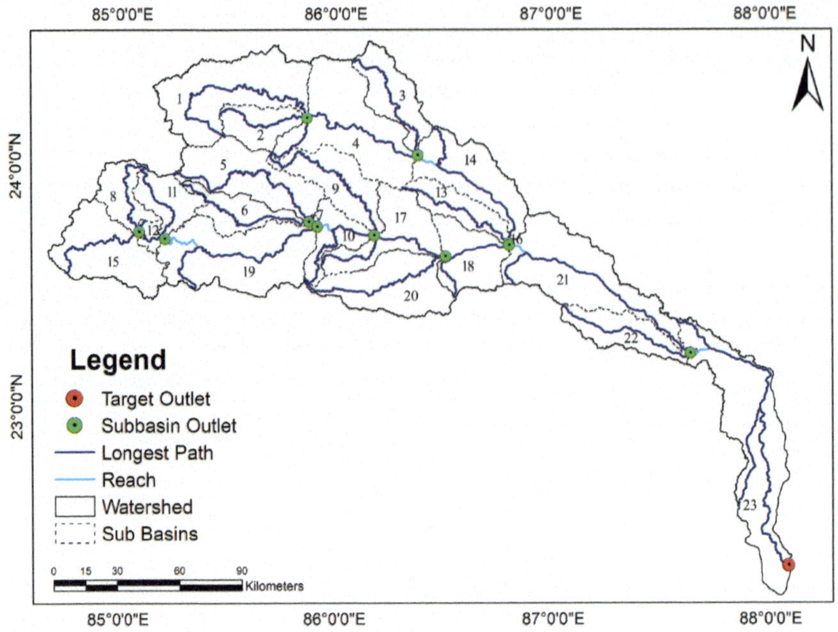

Fig. 3.1 Catchment outlet and target outlet established in Damodar River Basin (DRB) for SWAT model

directly impact fluvial sediment yield, surface runoff, and discharge. So, the variation of discharge, surface runoff, and sediment yield have been found in this catchment area. Thus, the model is applied to analyze the annual and monthly variations in discharge, sediment yield, and surface runoff at the outlet of each subbasin.

3.4.1 *Water Discharge Pattern*

3.4.1.1 Yearly Pattern

The SWAT model was used to simulate the annual discharge in the Damodar River Basin (DRB) from 1982 to 2012, with an average annual discharge of 2615.84 cumecs. The standard deviation of the mean annual discharge is 985.77 cumecs, resulting in a variability of 37.68%. The simulation for the post-dam period reveals a significant decrease in annual flow downstream over time, along with increased irregularity (Fig. 3.2). Notably, the years 1990, 1997, 2007, and 2011 recorded the highest mean annual flow, while 1982, 1992, and 2010 experienced the lowest flows. Over the 30 years, 14 years showed mean annual flows above the total average, while 18 years had flows below it. This variation indicates a substantial inconsistency in the annual flow of the Damodar River.

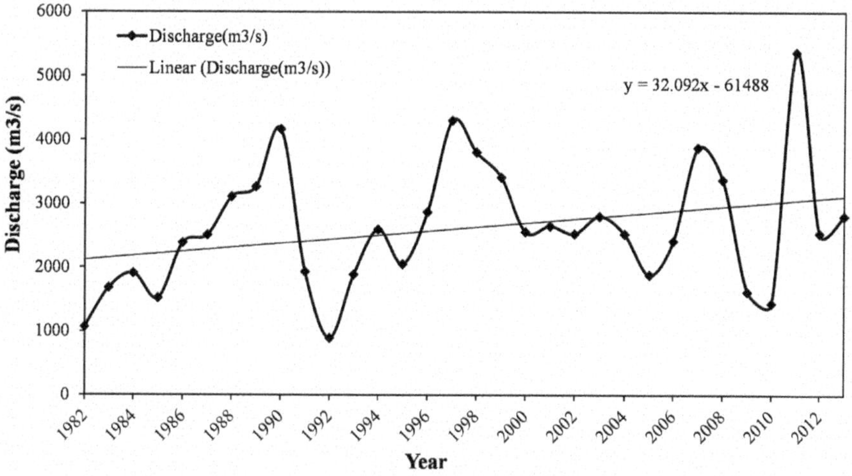

Fig. 3.2 Mean annual discharge of Damodar catchment during the period of 1982–2013

3.4.1.2 Monthly Pattern

The SWAT model was employed to simulate monthly discharge in the Damodar River Basin (DRB). The maximum mean monthly discharge occurs during the monsoon season, from June to September. The boxplot indicates that March, April, and May have the lowest average monthly flow within this catchment. Conversely, February, March, April, and May exhibit the highest variability in discharge throughout the year, while July, August, September, October, November, and December show the lowest variability in discharge in the Damodar catchment. The discharge is more consistent during the monsoon months and more variable during the pre-monsoon period (Fig. 3.3). The highest monthly average flow, recorded at 377.4 cumecs, occurred in August for the entire study period from 1982 to 2013. Overall, the monthly mean discharge for the entire period demonstrates a strong positive skew in this catchment.

3.4.2 Sediment Yield (SY) Pattern

3.4.2.1 Yearly Sediment Yield

Estimating sediment yield (SY) is crucial for understanding the fluvial morphology and dynamics within individual catchment areas. The SWAT model was utilized to assess sediment yield in the Damodar River Basin (DRB). During the period from 1982 to 2013, the Damodar catchment exhibited an increasing trend in sediment yield. The mean annual sediment yield for this period was 40.38 tons per hectare per year (t $ha^{-1}yr^{-1}$), with a standard deviation of 12.95 and a variability of 32.06%. The

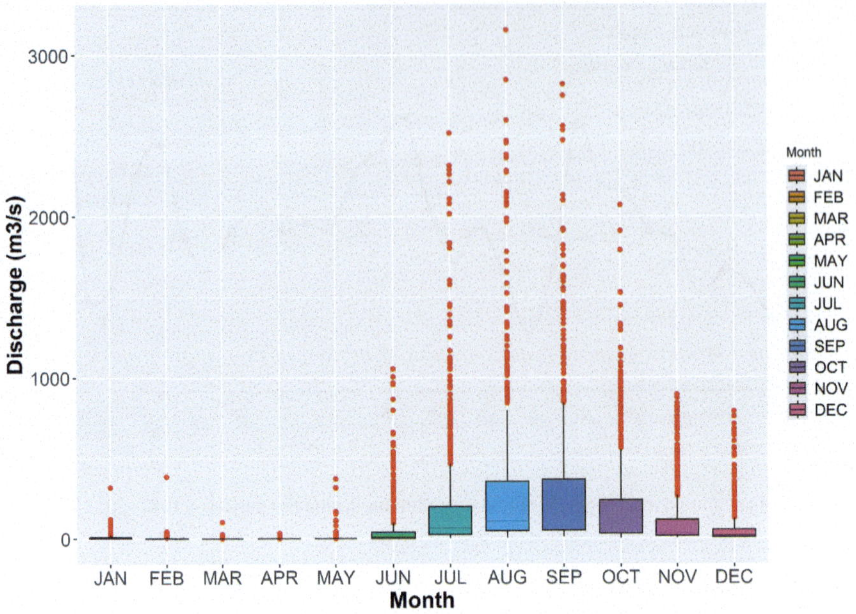

Fig. 3.3 Mean monthly discharge of Damodar catchment from 1982 to 2013

maximum sediment yields were recorded at 53.63, 53.68, 55.47, 65.76, 60.35, and 57.85 t $ha^{-1}yr^{-1}$ in the years 1988, 1989, 1996, 1997, 2003, and 2011, respectively. In contrast, the lowest sediment yields were observed at 17.33, 20.57, 21.80, and 28.60 t $ha^{-1}yr^{-1}$ in 1983, 1987, 2000, and 2009, respectively (Fig. 3.4).

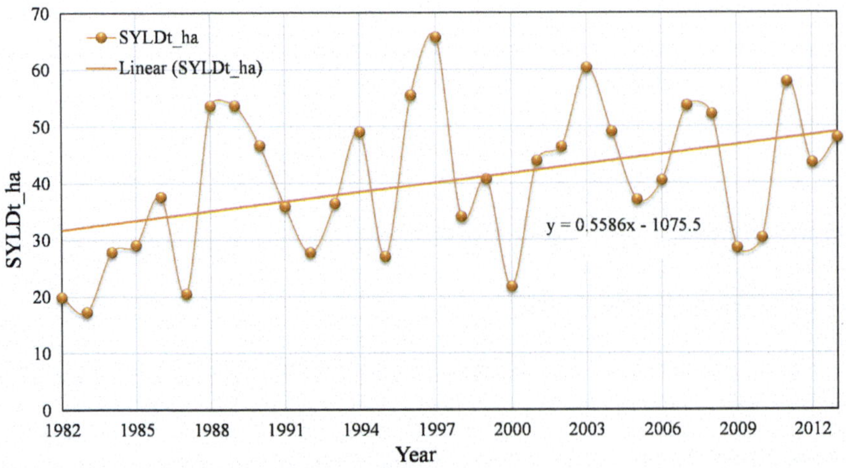

Fig. 3.4 Mean annual sediment yield (SY) of Damodar catchment during the period of 1982–2013

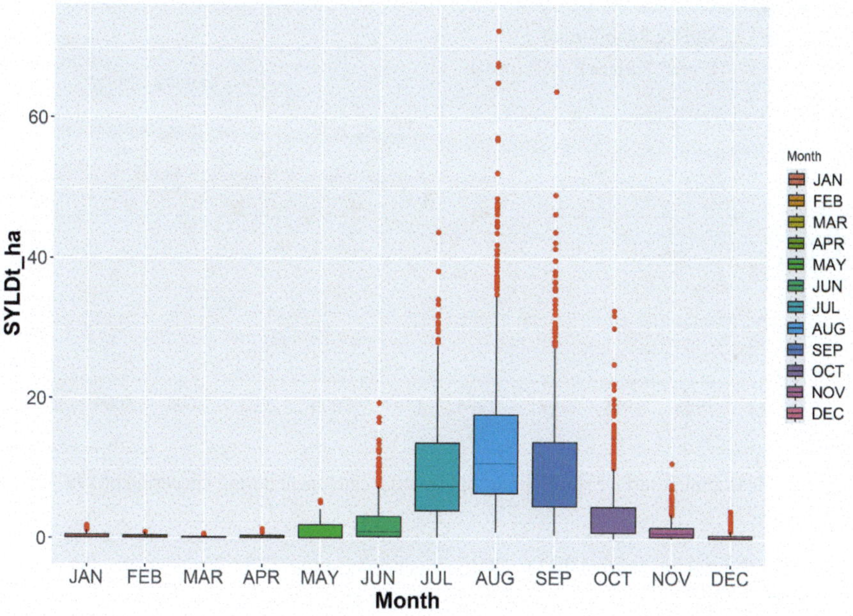

Fig. 3.5 Mean monthly sediment yield (SY) of Damodar catchment for the period of 1982–2013

3.4.2.2 Monthly Sediment Yield

The SWAT model was employed to simulate monthly sediment yield (SY) in the Damodar River Basin (DRB) for the period from 1982 to 2013, as illustrated in a boxplot (Fig. 3.5). The analysis indicates that months from February to May exhibit the highest variability in sediment yield, while the period from June to November displays the lowest variability. Notably, the maximum sediment yields occur in July, August, and September, with August recording the highest sediment yield for the entire study period. Furthermore, a positive correlation between sediment yield and fluvial discharge was observed. The monthly mean sediment yield for the entire duration reflects a moderately positive skewed distribution within the Damodar River Basin.

3.4.3 Surface Runoff Pattern

3.4.3.1 Yearly Runoff

Runoff is a critical component of the hydrological cycle that links precipitation with channel flow and groundwater flow. The SWAT model was utilized to simulate surface runoff (in mm) in the Damodar River Basin (DRB). The mean annual surface

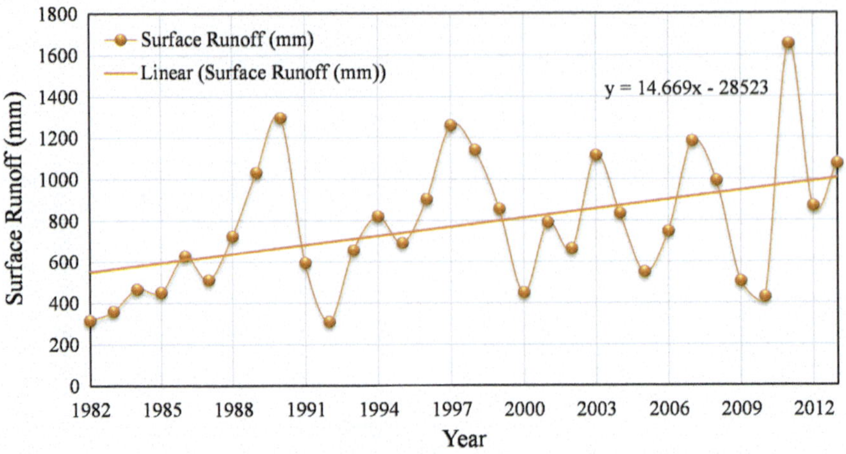

Fig. 3.6 Mean annual surface runoff (mm) of Damodar catchment during the period of 1982–2013

runoff for the period from 1982 to 2013 was recorded at 778.52 mm, with a standard deviation of 323.44 mm. The variability in mean annual runoff during this period was 41.54%. Out of the 32 years, more than 15 years exhibited runoff levels above the average. The maximum runoff values were observed in 1990 (1298.42 mm), 1997 (1260.47 mm), 2007 (1184.43 mm), and 2011 (1654.50 mm). Conversely, the lowest runoff levels were recorded in 1982 (317.15 mm), 1983 (361.21 mm), 1992 (311.67 mm), and 2010 (429.85 mm). A slight increasing trend in mean annual surface runoff was noted throughout the study period (1982–2013) in the Damodar catchment area (Fig. 3.6).

3.4.3.2 Monthly Distribution of Runoff

The SWAT model was used to simulate monthly mean surface runoff (in mm) in the Damodar catchment. The boxplot illustrating the mean monthly surface runoff from 1982 to 2013 is presented in Fig. 3.7. The monsoon period exhibits the highest monthly runoff, while the pre-monsoon period shows the lowest runoff in the Damodar catchment. Notably, May displays the greatest variability in surface runoff at 230.15 mm showing a high level of inconsistency, whereas the monsoon months collectively show the lowest variability (< 50 mm) and high consistency. Additionally, the months of September, October, and November demonstrate a normal distribution of surface runoff. Overall monthly distribution reflects a highly positive skewed pattern.

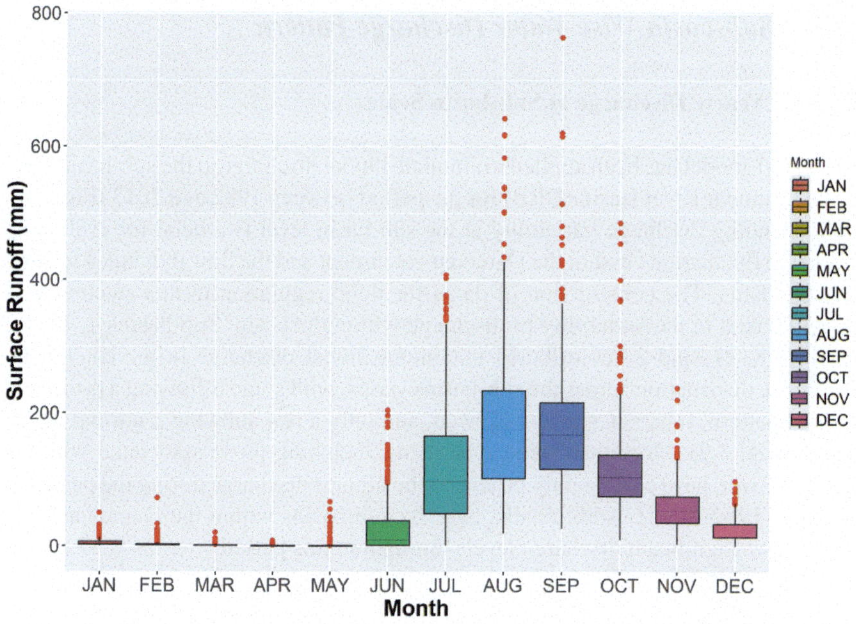

Fig. 3.7 Mean monthly surface runoff (mm) of Damodar catchment from 1982 to 2013

3.5 Water and Sediment Discharge: Study on Sub-basin of Damodar River Basin (DRB)

Sub-basin delineation within a GIS framework is contingent upon the selected threshold level, which dictates the number of sub-basins within a specific catchment area. In the Damodar catchment, 23 major sub-basins have been identified based on acceptable threshold levels. The SWAT model has been utilized to delineate these sub-basins and to simulate the monthly and yearly variations of discharge, sediment yield (SY), and surface runoff (in mm) at the sub-basin level. The dynamics of fluvial discharge, sediment yield, and surface runoff are significantly influenced by the factors of increasing population pressure, changing land use patterns, and mining activities. Furthermore, the construction of dams alters the fluvial system, while simultaneously exacerbating sedimentation issues that reduce trapping efficiency (TE). So, a detailed examination of the variations in discharge, sediment yield, and surface runoff at the sub-basin level is essential. This study provides an analysis of the yearly and monthly variations of these hydrological parameters for each of the sub-basins within the Damodar River Basin (DRB) catchment area.

3.5.1 *Sub-basin Wise Water Discharge Pattern*

3.5.1.1 Yearly Discharge at Sub-basin Scale

The SWAT model has been applied to simulate fluvial discharge at the sub-basin level in the Damodar River Basin (DRB) for the period between 1982 and 2013 (Fig. 3.8). Understanding discharge variability at the sub-basin level is crucial for evaluating discharge fluctuations within the Damodar catchment and the flow dynamics feeding into the dams. The construction of dams for flood regulation in this catchment is closely linked to the variability of discharge within the basin. Sub-basins 4, 10, 14, 16, 17, 18, 21, and 23 contributed maximum fluvial discharge in the catchment. Discharge distribution across the sub-basins varies, with some following a positively skewed pattern, others negatively skewed, and only a few showing a normal distribution (Fig. 3.9). More than three sub-basins, including those associated with the Barakar River, have consistently produced the highest discharge during the period of analysis (1982–2013). Additionally, over six sub-basins within the Damodar River have shown significant discharge levels throughout this period.

3.5.2 *Monthly Mean Discharge at Sub-basin Level*

The monthly mean discharge of 23 sub-basins was simulated to examine spatial variation from 1982 to 2013 (Fig. 3.10). A boxplot illustrates the monthly discharge variability across these sub-basins within the Damodar catchment. Sub-basins SB4,

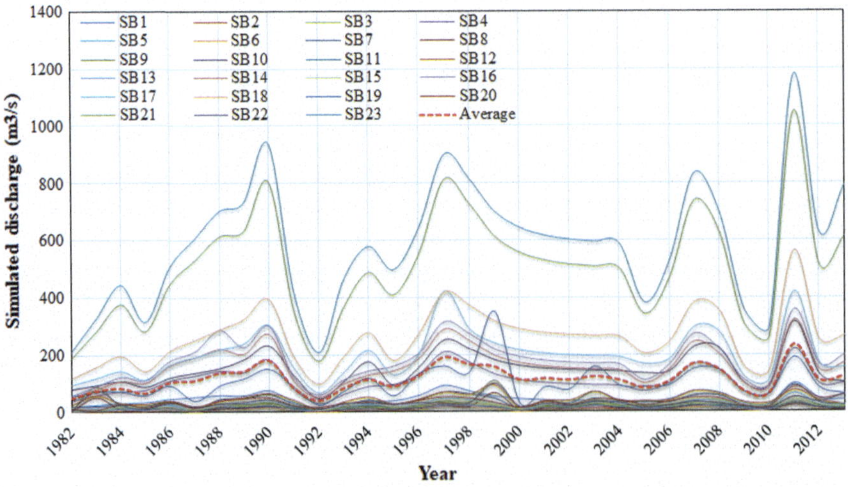

Fig. 3.8 Mean annual discharge of 23 sub-basins of the Damodar catchment for the period of 1982–2013

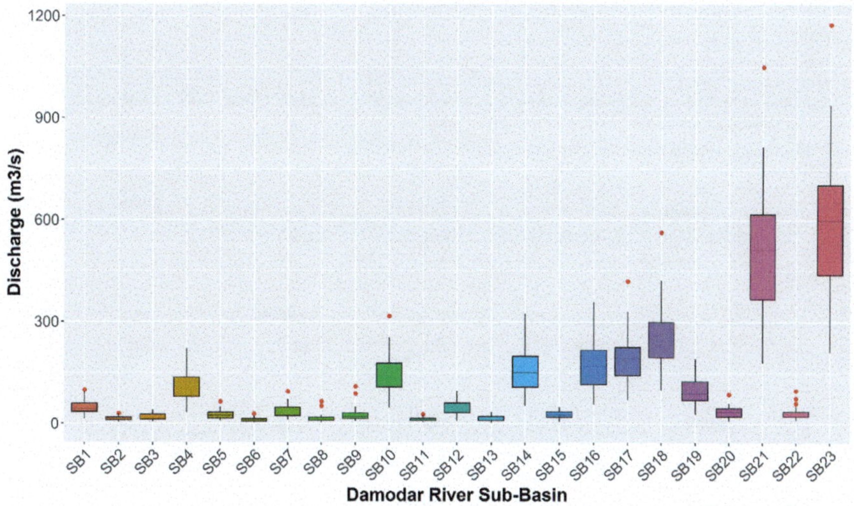

Fig. 3.9 The boxplot shows the distribution of the mean annual discharge of all sub-basins in the Damodar catchment from 1982 to 2013

SB10, SB14, SB16, SB17, SB18, SB21, and SB23 exhibit the highest monthly discharge, while the remaining sub-basins show lower discharge levels. The mean monthly discharge of sub-basins 21 and 23 are respectively high and the lowest mean monthly discharges over the entire period were recorded in SB2, SB6, and SB11, with values of 13.91, 11.40, and 13.15 m^3/s, respectively. Sub-basins located in the upper part of the Damodar catchment contribute minimum discharge to the main channel, whereas those in the middle section contribute significantly to the overall discharge output.

3.5.3 *Sediment Yield (SY) Pattern*

3.5.3.1 Yearly Sediment Yield at Sub-basin Level

Sediment plays a crucial role in understanding fluvial hydrodynamics and morphology. Sediment yield (SY) is influenced by various factors such as precipitation, geology, soil, vegetation, and surface runoff within a particular catchment area or outlet. In this study, the SWAT model was applied to simulate the annual mean sediment yield for all sub-basins (Fig. 3.11). Sub-basins SB6 and SB9 show the maximum average sediment yield, while SB15, SB22, and SB23 had the minimum yields due to their location on flat terrain (Fig. 3.12). Sub-basins SB2, SB4, SB6, SB7, SB9, SB14, and SB19 contribute significantly to sediment yield in the channel as they are situated in the upper and middle catchments of the Damodar basin, which is characterized by undulating and dissected terrain of the plateau region. The upper

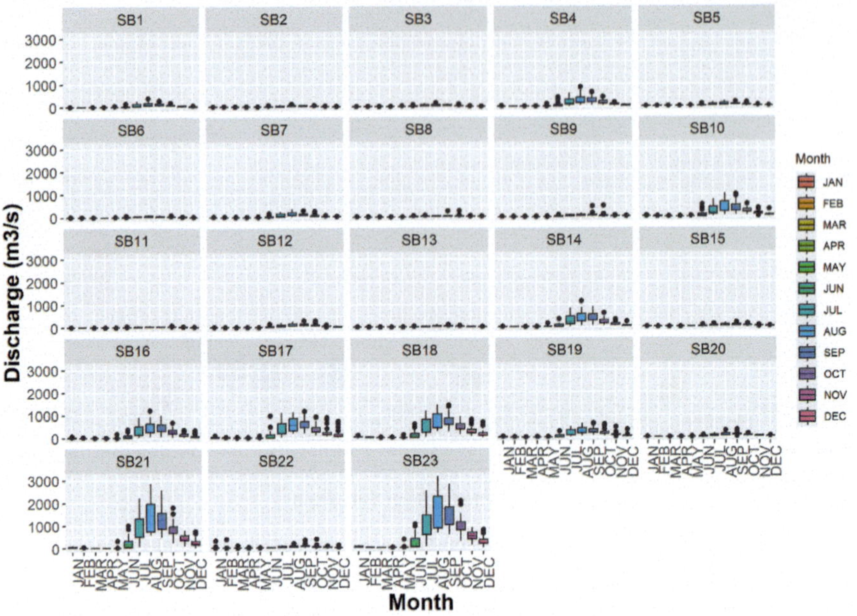

Fig. 3.10 The boxplot shows the distribution of the mean monthly discharge of all sub-basins in the Damodar catchment from 1982 to 2013

and middle catchment areas contribute the highest sediment yield in the Damodar Basin.

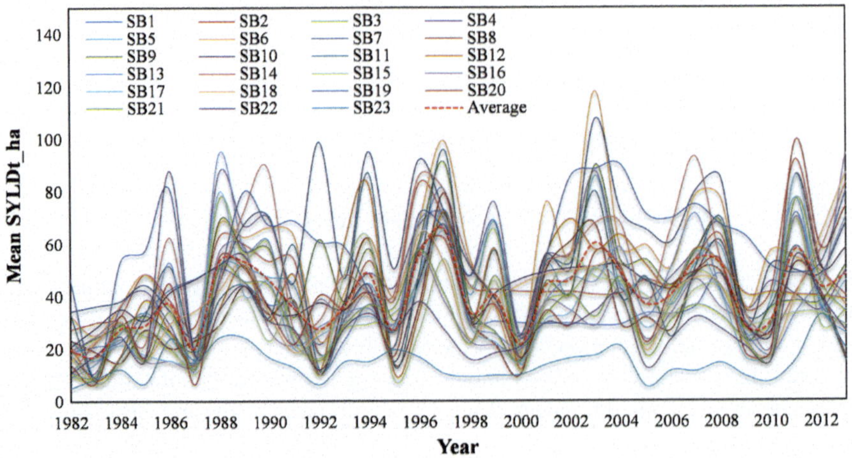

Fig. 3.11 Mean annual sediment yield (SY) of 23 sub-basins of the Damodar catchment in the period of 1982–2013

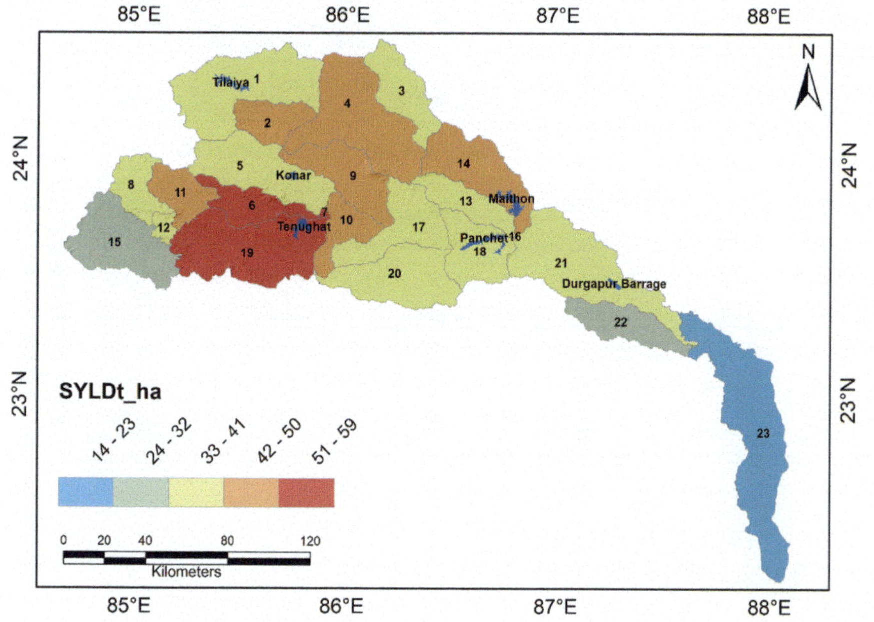

Fig. 3.12 The map shows the distribution of the mean sediment yield of all sub-basins in the Damodar catchment from 1982 to 2013

3.5.3.2 Monthly Sediment Yield at Sub-basin Level

The SWAT model was used to simulate the monthly sediment yield across various sub-basins in the Damodar catchment from 1982 to 2013 (Fig. 3.13). A boxplot illustrates the variation in the monthly mean sediment yield (SY) across the sub-basins. The upper and middle catchments of the Damodar Basin host major dams, including Panchet, Maithon, Tenughat, Konar, and Tilaya. The upper and middle catchments of the Damodar Basin produce maximum sediment yield despite low discharge due to factors like highly erodible soils, steep slopes, sparse vegetation, and intense mining activities.

3.5.4 Surface Runoff Pattern

3.5.4.1 Yearly Surface Runoff at Sub-basin Scale

Surface runoff is a critical hydrological variable in catchments, representing the integrated effect of precipitation and channel flow. Its intensity is closely linked to sediment yield, with elevated runoff corresponding to increased sediment transport into the channel, and vice versa. The SWAT model was employed to examine the spatial

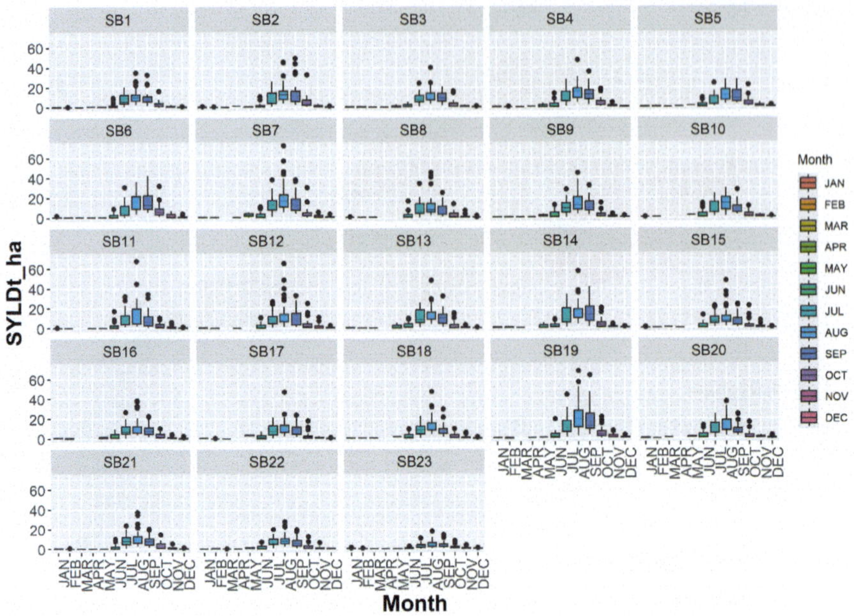

Fig. 3.13 The boxplot shows the distribution of mean monthly sediment yield (SY) of all the sub-basins in the Damodar catchment from 1982 to 2013

and temporal variation of annual mean surface runoff across sub-basins within the Damodar catchment. The analysis revealed that the northwestern and lower regions of the basin exhibited the maximum runoff levels, while the southwestern region recorded the lowest during the period 1982–2013 (Fig. 3.14). Notably, the years 1990, 1997, 2003, 2007, and 2011 experienced peak surface runoff across all sub-basins (Fig. 3.15), contrasting with the significantly lower runoff observed in 1982, 1987, 1992, 2000, and 2010. Over the study period, nine sub-basins registered an average surface runoff exceeding 750 mm, whereas 14 sub-basins recorded averages below this threshold, highlighting the spatial variability in runoff dynamics within the Damodar catchment.

3.5.4.2 Monthly Pattern of Surface Runoff at Sub-basin Level

The sub-basin-wise monthly variation in surface runoff (mm) is crucial for effective dam management in the Damodar catchment. The SWAT model was used to compute the mean monthly surface runoff for each sub-basin. Boxplots reveal the variation in monthly runoff, with sub-basins SB16, SB18, SB21, SB22, and SB23 contributing the highest runoff to the channel (Fig. 3.16). The majority of surface runoff occurs during the monsoon season (June to September), with a highly skewed distribution during this period. In contrast, the pre-monsoon season exhibits significantly lower

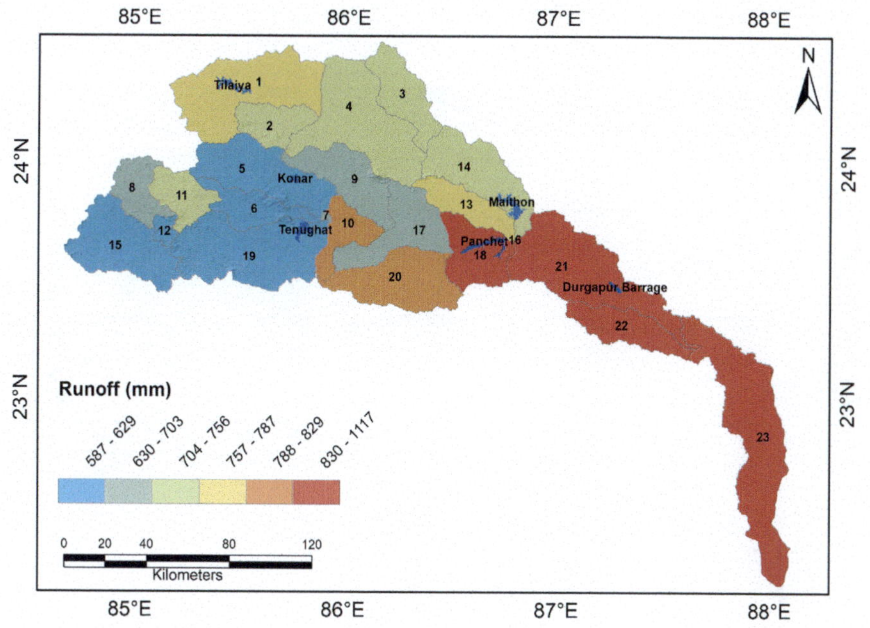

Fig. 3.14 The annual surface runoff (mm) of 23 sub-basins of the Damodar catchment in the period of 1982–2013

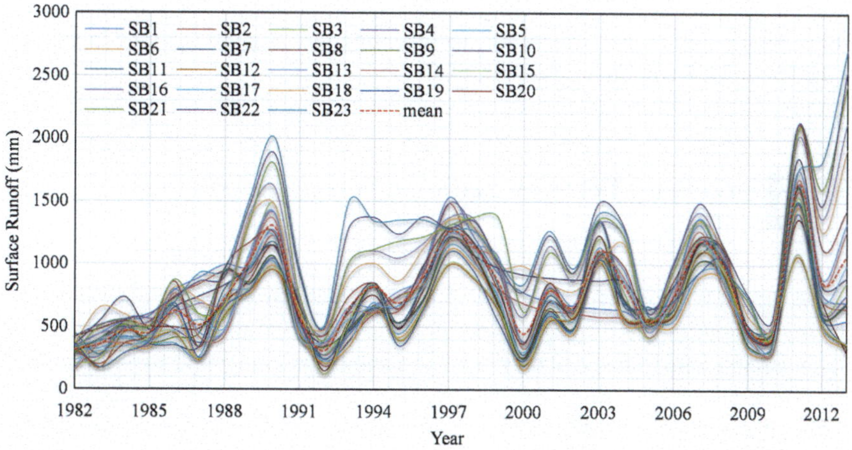

Fig. 3.15 The map shows the distribution of mean surface runoff (mm) of all the sub-basins in the Damodar catchment from 1982 to 2013

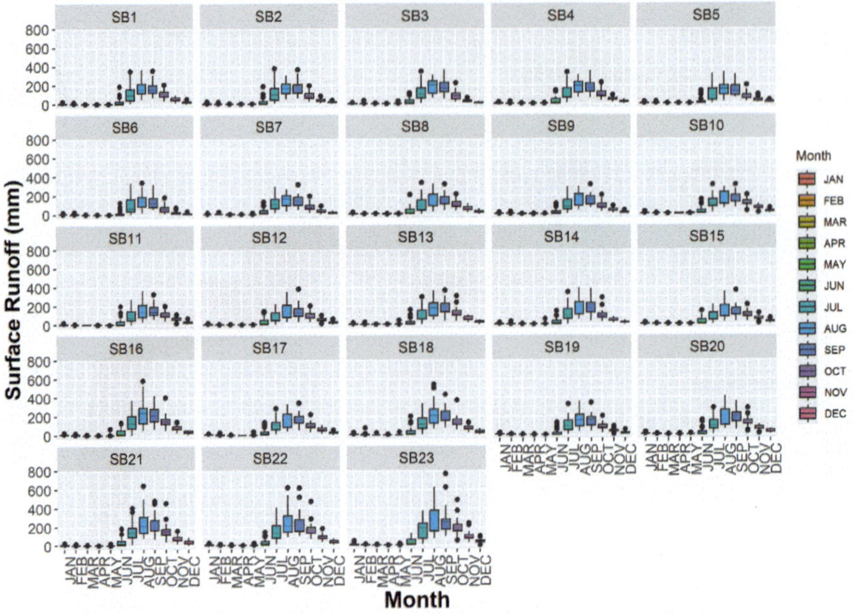

Fig. 3.16 The boxplot shows the distribution of mean monthly surface runoff (mm) of all the sub-basins in the Damodar catchment from 1982 to 2013

runoff, along with higher variability, underscoring the strong seasonal dependency of the Damodar River on monsoon rainfall.

3.6 Relationship Between Sediment Yield and Surface Runoff

Sediment yield and surface runoff are pivotal variables in the Damodar catchment for understanding fluvial morphodynamics and sediment input to dam reservoirs. These variables are influenced by a range of factors, including land cover, precipitation, temperature, vegetation, soil characteristics, and geology. Importantly, sediment yield in the channel is predominantly determined by the intensity of surface runoff. Thus, elucidating the relationship between these two variables is vital for effective dam management. In this analysis, surface runoff is treated as the independent variable, while sediment yield serves as the dependent variable. The relationship exhibits a positive correlation, with an R^2 value of 0.65, indicating that 65% of the variability in sediment yield can be attributed to surface runoff (Fig. 3.17b). Based on a comprehensive dataset spanning 32 years, the regression equation established is: sediment yield (y) = 0.0323 * surface runoff (x) + 15.234. Notably, maximum sediment yields associated with elevated surface runoff were recorded in 1997 and

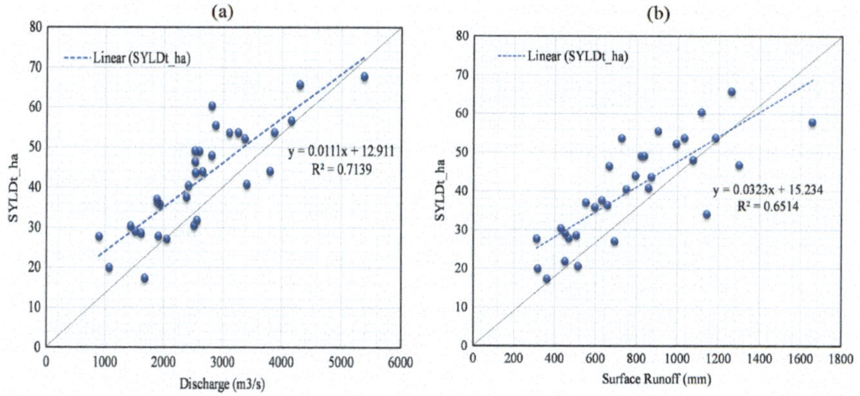

Fig. 3.17 The scatter plot shows (a) a positive relationship between discharge and sediment yield of the Damodar basin, and (b) a positive relationship between surface runoff and sediment yield of the Damodar basin

2011. This positive correlation highlights the necessity for effective management of surface runoff in the Damodar basin, as it indirectly influences sediment yield delivered to the channel. Sub-basins of high surface runoff are to be given more priority in soil erosion management.

3.7 Relationship Between Water Discharge and Sediment Yield

Fluvial discharge and sediment yield exhibit a positive correlation, as demonstrated by the results of the SWAT model simulations conducted for the Damodar basin. In this study, a comprehensive analysis of 32 years of mean data was performed to elucidate the dynamics between these two critical variables within the catchment (Fig. 3.17a). The analysis yielded an R^2 value of 0.71, indicating that fluvial discharge accounts for 71% of the variability in sediment yield in the Damodar catchment. The established regression equation is as follows: sediment yield (y) = 0.0111 * discharge (x) + 12.911 (Table 3.1). The topography of the Damodar catchment, characterized by an amalgamation of plateaus and hills, combined with the predominantly flat regions in the lower catchment, influences the positive relationship between sediment yield and discharge across these areas. Consequently, substantial sediment inputs are generated within the channels and various dams. Therefore, it is imperative to implement effective management strategies for both discharge and sediment yield to mitigate siltation challenges faced by multiple dams in the Damodar catchment.

Table 3.1 Linear regression statistics of 1. Relationship with discharge and sediment yield and 2. Relationship with surface runoff and sediment yield (SY)

Relationship with variables	Regression statistics		Regression equation	Interpretation
1. Discharge versus sediment yield (SY)	Multiple R	0.8449	Sediment yield (y) = 0.0111 discharge (x) + 12.911	This is significantly correlated (p-value < 0.001) with each other
	R square	0.7139		
	Adjusted R square	0.7044		
	Standard error	7.0408		
	Observations	32		
2. Surface runoff versus sediment yield (SY)	Multiple R	0.8071	sediment yield (y) = 0.0323 surface runoff (x) + 15.234	This is significantly (p-value < 0.0002) with each other
	R square	0.6514		
	Adjusted R square	0.6398		
	Standard error	7.7697		
	Observations	32		

Relationship with discharge and sediment yield and 2. Relationship with surface runoff and sediment yield (SY)

3.8 SWAT Model Validation

Model validation is a critical component of this study, involving the assessment of simulated results against observational data. A model is considered valid when its predictions align satisfactorily with empirical observations. In this research, the SWAT model has been employed to simulate discharge, sediment yield, and surface runoff within the Damodar catchment. Validation processes were specifically conducted for fluvial discharge and surface runoff. To achieve this, the simulated discharge and surface runoff were compared with observed data obtained from various sources, including research articles, theses, books, and secondary information. The model's performance was evaluated using several statistical measures, including R^2, adjusted R^2, root mean square error (RMSE), and mean absolute deviation (MAD). All statistical indicators fell within acceptable ranges, thereby confirming the robustness and reliability of the model's results.

3.8.1 Water Discharge

The simulated discharge data for the Damodar River basin from 1981 to 2013 were validated using observed data obtained from the Damodar Valley Corporation (DVC). The performance of the model was assessed by comparing simulated discharge with observed discharge using various statistical measures (Table 3.2). Figure 3.18 illustrates the comparison between simulated and observed discharge for the Damodar basin from 1981 to 2007. Notably, the years 1984, 1985, 1994, 1995, and 2007 exhibited underestimations in the simulated discharge, while the years 1982, 1997, and 2004 showed instances of overestimation. The analysis reveals a strong agreement between the simulated and observed discharge values, as evidenced by the R2 value of 0.87 (Fig. 3.19), indicating that 87% of the variability in discharge is accurately predicted by the SWAT model. The root mean square error (RMSE), a standard metric for quantifying model prediction error, is recorded at 132.53 m^3/s. Additionally, the mean absolute deviation (MAD), a robust measure of variability in quantitative data, is found to be 90.85 m^3/s (Table 3.2). These statistical indicators collectively affirm the reliability and effectiveness of the SWAT model in simulating discharge within the Damodar River basin.

Table 3.2 Various statistical measures to compare simulated and observed data of the Damodar catchment (for discharge) and Hazaribagh, Bokaro, and Dhanbad districts (for surface runoff)

Validation statistics				
Statistical measured	Discharge	Surface runoff		
		Hazaribagh	Bokaro	Dhanbad
R^2	0.87	0.78	0.89	0.91
Adjusted R^2	0.86	0.71	0.85	0.88
RMSE	132.53	107.52	122.13	73.45
MAD	90.85	84.41	111.65	62.98

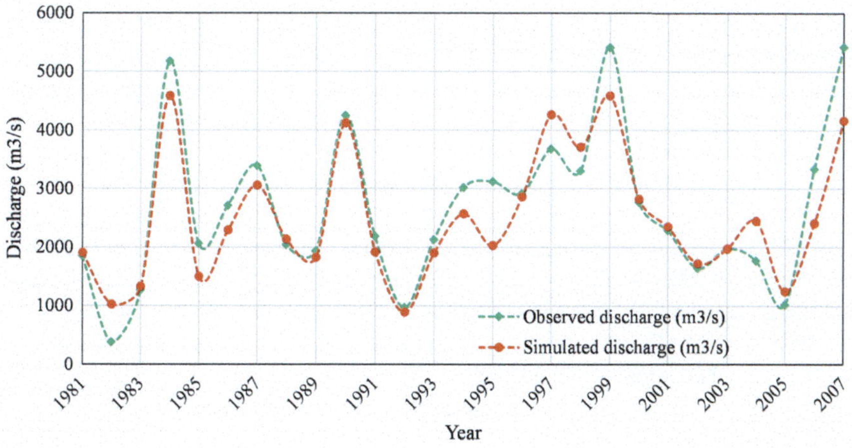

Fig. 3.18 Comparison between observed and simulated yearly mean discharge of Damodar basin from 1981 to 2007

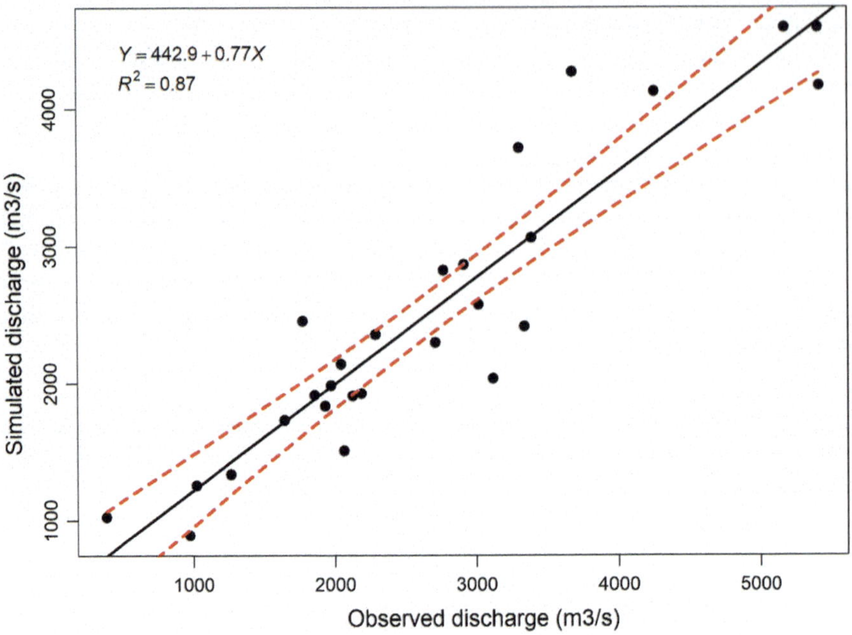

Fig. 3.19 Scatter plot of yearly mean discharge in Damodar basin (1981–2007)

3.8.2 *Surface Runoff*

The SWAT model was employed to simulate surface runoff in the Damodar catchment, with validation data sourced from the article by [4]. The simulated results were compared against observed surface runoff data from Hazaribagh, Bokaro, and Dhanbad for the period from 2006 to 2010. The R^2 values for these districts were 0.78, 0.89, and 0.91 for Hazaribagh, Bokaro, and Dhanbad, respectively, indicating a strong correlation between simulated and observed surface runoff. The root mean square error (RMSE) and mean absolute deviation (MAD) values are presented in Table 3.2. The minimum RMSE value of 73.45 mm was recorded in Dhanbad, while Bokaro and Hazaribagh had RMSE values of 122.13 and 107.52 mm, respectively (Fig. 3.20). The minimum MAD value was also found in Dhanbad at 62.98 mm, compared to 84.41 mm in Hazaribagh and 111.65 mm in Bokaro. Overall, the accuracy of the simulations for Hazaribagh, Bokaro, and Dhanbad was satisfactory.

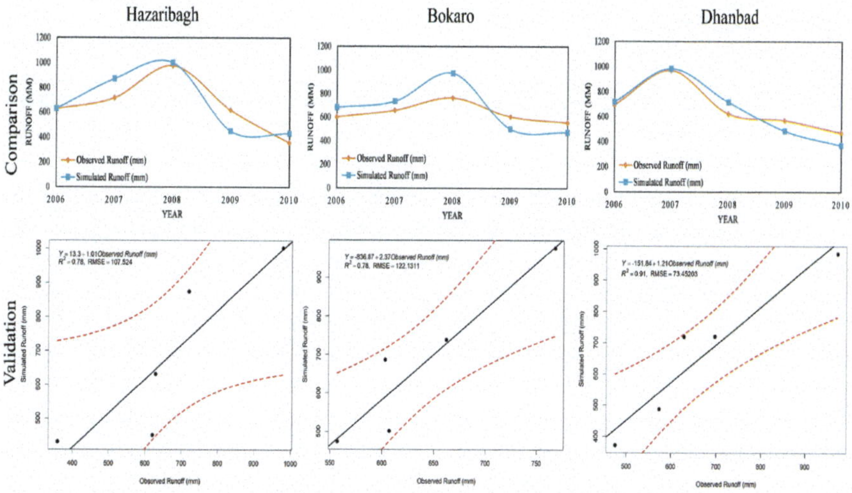

Fig. 3.20 The graph compares simulated and observed runoff and scatter plots, indicating the validation of the Hazaribagh, Bokaro, and Dhanbad districts

3.9 Major Findings

The summary of various findings serves to guide decision-making for future management plans for the Damodar catchment. The key findings of this chapter are as follows:

- The annual mean discharge, sediment yield (SY), and surface runoff for the entire Damodar catchment during the period from 1982 to 2013 are 2615.84 m^3/s, 40.38 t $ha^{-1}yr^{-1}$, and 778.52 mm, respectively.
- The monthly mean discharge, sediment yield, and surface runoff peak in August and September, exhibit low variability. In contrast, the pre-monsoon months demonstrate the lowest average values and maximum variability throughout the entire period.
- The upper and middle catchments of the Damodar Basin produce maximum sediment yield despite low discharge due to factors like highly erodible soils, steep slopes, sparse vegetation, and intense mining activities.
- The lower catchments of the Damodar River basin (DRB) are characterized by the highest discharge and surface runoff throughout the entire period from 1982 to 2013, whereas the upper catchment of the Damodar basin exhibits the greatest sediment yield.
- It is noted that sub-basins SB4, SB6, SB14, and SB19 yield the highest average outputs for all three variables (discharge, sediment yield, and surface runoff) due to their location on the plateau in the upper and middle catchment. These sub-basins are identified as highly vulnerable to sediment yield.

3.10 Conclusion

In this chapter, the SWAT model is employed to derive hydrometeorological data for ungauged large Damodar catchments for the duration of 1981–2013. The Damodar River Basin (DRB) has been significantly influenced by human modifications and alterations, particularly through open-cast mining and the establishment of dams, which have disrupted the natural fluvial hydrometeorological system of the river. The upper and middle catchments of the Damodar basin are characterized by substantial sediment production, contributing significantly to channel inputs. The SWAT model has generated monthly and yearly data on discharge, surface runoff, and sediment yield (SY) to analyze the variability within the Damodar basin. These three variables have been calculated at the sub-basin level to identify key locations that affect variability and impact dam performance. As a result, several highly productive sub-basins have been identified as crucial sources of discharge, surface runoff, and sediment input to the main channel, thus influencing dam operations.

In summary, all simulated data have been validated using various statistical measures, yielding results within acceptable ranges. The findings of this study identify the sub-basins that generate the highest outputs of discharge, surface runoff, and sediment yield to the main channel. This chapter delineates the mean monthly and yearly spatial variability of hydrometeorological factors across different sub-basins of the Damodar catchment.

References

1. Arnold JG, Srinivasan R, Muttiah RS, Williams JR (1998) Large area hydrological modeling and assessment. Part I: model development. J Am Water Resour Assoc 34
2. Charlton R (2007) Fundamentals of fluvial geomorphology. Routledge
3. Dile YT, Daggupati P, George C et al (2016) Introducing a new open source GIS user interface for the SWAT model. Environ Model Softw 85:129–138. https://doi.org/10.1016/j.envsoft.2016.08.004
4. Ghosh S, Mistri B (2015) Geographic concerns on flood climate and flood hydrology in Monsoon-Dominated Damodar River Basin, Eastern India. Geogr J 2015:1–16. https://doi.org/10.1155/2015/486740
5. Hajigholizadeh M, Melesse AM, Fuentes HR (2018) Erosion and sediment transport modelling in shallow waters: a review on approaches, models and applications. Int J Environ Res Public Health 15. https://doi.org/10.3390/ijerph15030518
6. Mutchler CK, Murphree CE, McGregor KC (1994) Laboratory and field plots for erosion research. In: Soil and Water Conservation Society (U.S.) (ed) Soil erosion research methods. Taylor and Francis Ltd., p 28
7. Neitsch SL, Arnold JG, Kiniry JA, Williams JR (2009) Soil and water assessment tool theoretical documentation version 2009. SWAT 618. https://doi.org/10.1016/j.scitotenv.2015.11.063
8. Roy P, Roy A, Joshi P et al (2015) Development of decadal (1985–1995–2005) land use and land cover database for India. Remote Sens 7:2401–2430. https://doi.org/10.3390/rs70302401
9. Schürz C (2019) Development of tools and interfaces for the implementation, pre-processing, and sensitivity analysis of the SWAT model in an R programming environment Report to the Austrian Marshall Plan Foundation

10. Williams JR (1975) Sediment routing for agricultural watersheds 1. JAWRA J Am Water Resour Assoc 11:965–974. https://doi.org/10.1111/j.1752-1688.1975.tb01817.x
11. Wischmeier WH, Smith DD (1978) USLE Predicting rainfall erosion lossses: a guide to conservation planning
12. Wu Y, Liu S, Abdul-Aziz OI (2012) Hydrological effects of the increased CO_2 and climate change in the Upper Mississippi River basin using a modified SWAT. Clim Change 110:977–1003. https://doi.org/10.1007/s10584-011-0087-8

Chapter 4
Identifying Soil Erosion Hotspot: Sediment Delivery Ratio Modelling

Abstract This chapter focuses on identifying soil erosion hotspots (SEH) in the Damodar River Basin, specifically within the catchments of Maithon, Panchet, and Tenughat Dams, using Sediment Delivery Ratio (SDR) modeling. The objective is to examine the relationship between soil erosion (SE) and sediment yield (SY) and understand how erosion hotspots influence sedimentation in the dam catchments. SDR modeling helps quantify the amount of eroded soil that is transported to water bodies, identifying critical areas with high erosion rates that contribute significantly to sediment yield. This chapter analyzes spatial variations in soil erosion, linking areas of intense erosion to increased sediment delivery and deposition in the dams. The impact of these hotspots on dam performance, including sedimentation and reservoir capacity, is assessed. Findings indicate that soil erosion hotspots are major contributors to sedimentation issues in the catchments, affecting the long-term sustainability and efficiency of the dams. The research underscores the importance of identifying and managing erosion hotspots to mitigate sedimentation, improve soil conservation practices, and enhance reservoir management strategies in the region. This chapter provides valuable insights for developing targeted interventions to reduce soil erosion and its adverse effects on sedimentation in the Damodar River Basin's major dams.

Keywords Soil erosion · RUSLE model · Sediment delivery ratio · Soil erosion hotspot · Slope length · Soil erodibility · Normalized difference vegetation index

4.1 Introduction

Globally, land degradation is a significant issue [8]. Increasing population pressure on land and development activities poses considerable hazards to the land, which is plagued by severe soil erosion (SE) and land degradation [57]. Land degradation or soil erosion (SE) is a momentous threat to the world's ecosystems or inhabitants [6, 23, 45]. India is one of the fastest-growing countries in the world and it is confronted with the maximum developmental activities, such as mining, agricultural activities, deforestation, etc. Damodar basin is known for coal deposits which account for 46%

S. A. Alam and R. Maiti, *Sediment Source to Sink: Deciphering Sediment Connectivity to Large Dams in Damodar River Basin*, Earth and Environmental Sciences Library,
https://doi.org/10.1007/978-3-031-87072-9_4

of the country's coal reserves [40]. Intensive mining, deforestation, and crop cultivation in the Damodar basin have triggered high-vulnerability soil erosion hotspot (SEH) zones which produce enormous sediment yield (SY) input to the main channel and various dams (Maithon, Panchet, Tenughat, Konar, and Tilaya). Therefore, the identification of spatial soil erosion (SE), sediment delivery ratio (SDR), and soil erosion hotspots (SEH) are essential for the management of land degradation problems and siltation problems in various dams of the Damodar River Basin (DRB). Various researchers have worked on the Damodar River basin which is related to soil erosion, sediment yield, etc. [7, 30, 33, 46, 47, 50]. The soil erosion hotspots (SEH) are the major source of contributing sediment to both the main channel and dams. In this study, we developed a modeling framework to assess the impact of varying soil erosion (SE) severity, sediment delivery ratio (SDR), and soil erosion hotspots (SEH) on dams within the Damodar River Basin (DRB).

4.2 Conceptual Background of Modelling: Soil Erosion (SE), Sediment Delivery Ratio (SDR), and Soil Erosion Hotspot (SEH)

Over the past several decades (1990–2021), comprehensive models for soil erosion (SE), sediment yield (SY), sediment delivery ratio (SDR), and soil erosion hotspots (SEH) have been developed. However, the increasing complexity of natural processes due to ongoing development in recent years has posed significant challenges to modeling these phenomena. To address this, we employ conceptual models to better understand the diversity, complexity, and intricate interactions between SY, SDR, and SHE.

4.2.1 Soil Erosion Rate (SE)

Human interference with land is continuously increasing, which in turn accelerates land degradation [8]. The major quantity of soil is displaced by human activities which is responsible for the soil erosion process [27]. The extent of soil erosion worldwide is mostly caused by a variety of activities, including mining, building, deforestation, and agriculture. More than 6 billion hectares of areas are influenced by various land degradation activities [23, 58]. Soil erosion has accelerated over time due to changes in various factors and their interactions. In India, over 45% of land is severely impacted by soil erosion, primarily affecting agricultural areas and contributing to siltation in reservoirs [31]. Damodar River Basin (DRB) is highly affected by severe soil erosion due to various human encroachments which act as major sediment input to the channels and dams. Therefore, the examination of spatial soil erosion (SE) patterns in the Damodar basin is required for management planning. Numerous soil erosion models are utilized globally, including the Universal Soil Loss Equation (USLE) developed by [60, 61]), as well as its updated version, the Revised Universal Soil Loss Equation (RUSLE) [48]; Water Erosion Prediction Project (WEPP) [43]; and the European Soil Erosion Model (EUROSEM) [42], etc. In this study, we employed the RUSLE model because of its simplicity, low data requirements, easy access to input data, and suitability for application in large basins. Scholars worldwide have utilized the RUSLE model under various conditions [8, 10, 12, 19, 23, 30, 57, 62, 65]. Therefore, identifying soil erosion severity zones is essential for addressing siltation issues in the dams within the Damodar River Basin (DRB).

4.2.2 Sediment Delivery Ratio (SDR)

Sediment production in a watershed mainly results from various geological and hydrological processes, including weathering, erosion, surface runoff, fluvial activities, and human impacts. This sediment is then transported through the channel to the catchment outlet. The estimation of sediment yield to the basin is incorporated with the delivery process of sediments of that particular basin. In large watersheds, a significant amount of sediment is deposited within the watershed itself, with only a small percentage of the soil eroded from hillslopes reaching the stream and the watershed outlet [18, 36]. Therefore, the sediment delivery ratio (SDR) is significant for the estimation of actual sediment yield to the catchment. SDR expresses the eroded sediment delivery probability of reaching the nearest stream channels or outlets from a certain upland location [21]. Several researchers have developed methods to estimate the sediment delivery ratio (SDR) using different algorithms and rainfall-runoff factors [2], relief-length ratio [38], particle size [56], water runoff travel time [21], and runoff curve numbers [59]. In India, various scholars have worked on the sediment delivery process [1, 28, 30, 53]. The Damodar River Basin (DRB) is significantly impacted by severe soil erosion resulting from various human encroachments, which contribute major sediment inputs to the channels and dams. Therefore, estimating the sediment delivery process in each sub-watershed is crucial for understanding the nature of sediment yield for sustainable management and planning.

4.2.3 Soil Erosion Hotspot (SEH)

A hotspot refers to a spatial analysis technique that detects statistically significant clustering patterns of high values (hot spots) and low values (cold spots). In this study, we utilized the Getis-Ord Gi* statistics to identify soil erosion hotspots (SEH) within the Damodar River Basin (DRB). These methods find the spatial association among neighboring sample points and identify the statistically significant spatial clusters of high and low point values as hotspots and coldspots, respectively [24]. The outputs of the Getis-Ord Gi* statistics generate new feature classes, including Z-score, *P*-value, and Confidence Level (CL). Features exhibiting a high Z-score and a small *P*-value are indicative of statistically significant hotspots, while features with a low Z-score and a small *P*-value signify statistically significant cold spots. Hotspot analysis for soil erosion has been used by several researchers to identify important erosion-impacted locations. Hotspot analysis has been effectively utilized in conjunction with soil erosion, ecosystem services, deforestation, and biodiversity studies by various scholars worldwide [9, 11, 16, 34, 35, 37, 44]. Identification and mapping of the soil erosion hotspots (SEH) in the Damodar catchment seems to be essential for targeted policy making.

4.3 Estimation of Soil Erosion Rate

Soil erosion is a significant problem all over the world [8]. The selection of the soil erosion prediction model depends on the location, size, and climatic condition of this study area. In this study, we employed the RUSLE model because of its universal acceptability, simplicity, limited data requirements, easy access to input data, and suitability for application in large basins. Scholars worldwide have utilized the RUSLE model under various conditions [8, 10, 12, 19, 23, 30, 57, 62, 65].

4.3.1 *RUSLE Model*

The Revised Universal Soil Loss Equation (RUSLE) model is based on an empirical equation that predicts annual soil loss for a specific region, incorporating five key variables [48]. The model inputs consist of five independent factors: Rainfall erosivity (R), Soil erodibility (K), Slope length and slope steepness (LS), Cover management (C), and Support practice (P). These factors are utilized in the RUSLE model, as shown in Eq. (4.1). The model operates within a GIS environment and is calculated on a cell-to-cell basis. The RUSLE method is expressed as following Eq. (4.1).

$$SE = R_{i,j} * K_{i,j} * LS_{i,j} * C_{i,j} * P_{i,j} \tag{4.1}$$

where,

SE Annual soil loss (t ha^{-1}yr^{-1})
$R_{i,j}$ Rainfall erosivity factor (MJ mm h^{-1} ha^{-1} year^{-1})
$K_{i,j}$ Soil erodibility factor (t ha h ha^{-1}MJ^{-1} mm^{-1})
$LS_{i,j}$ Slope length factor (Unit less)
$C_{i,j}$ Land coverage factor (Unit less)
$P_{i,j}$ Support practice factor (Unit less).

4.3.2 *Factors of the RUSLE Model*

4.3.2.1 Rainfall Erosivity Factor (R-Factor)

The rainfall erosivity factor is an important component in the RUSLE-GIS model because it predicts the severity of soil erosion in a given area. This component quantifies rainfall's erosive strength, which is determined by its volume, intensity, and duration. The potential soil erosion by the power of rainfall is called rainfall erosivity [61]. According to [60] and [48], the estimation of the R factor requires total rainfall energy (E) and a maximum 30-min rainfall intensity (I30) of storm events for at least 20 years. Such high-resolution data are not available in many countries, including India. In this context, Scholars in India have developed their empirical formulas, which are applied in various regions across the country [3, 51]. In this research, we have gathered 40 years of annual rainfall data (1980–2020) from the Indian Meteorological Department (IMD) to calculate the rainfall erosivity factor using the equation formulated by [3].

$$R = 81.5 + 0.38(R_N) \tag{4.2}$$

For,

$$340 \le R_N \le 3500 \text{ mm}$$

where, R_N Corresponding to mean annual rainfall in mm, R expressed annual erosivity index (MJ mm h^{-1} ha^{-1} $year^{-1}$)

For this instance, we have selected 28 rain gauge stations around the Damodar River Basin (DRB) and acquired daily rainfall data from IMD for the estimation of the rainfall erosivity factor. Further, the average annual rainfall (mm) is calculated for each rain gauge station by using previously collected daily rainfall (mm) data. To estimate the R factor, we used Eq. (4.2), and the spatial distribution of R factors is generated by using the Kriging interpolation in the ArcGIS 10.3 environment. In this study, we created three R factor maps for the years 1990, 2005, and 2021 to illustrate the spatial distribution of the R factor in the study area as shown in Fig. 4.1. The estimated R factor values ranged from 490.28 to 697.56 MJ mm h^{-1} ha^{-1} $year^{-1}$ in 1990, while in 2005, they varied from 403.63 to 719.73 MJ mm h^{-1} ha^{-1} $year^{-1}$, and in 2021, the values were between 424.01 and 614.39 MJ mm h^{-1} ha^{-1} $year^{-1}$. In 1990, the lower catchment and the northwestern part of the drainage basin (DRB) had a significant R erosivity factor, while the southwestern region displayed the lowest. By 2005, the lower and middle-lower catchment areas had the highest rainfall erosive intensity, while the upper and middle-upper catchments recorded the lowest. In 2021, the lower catchment continued to display the highest erosivity intensity, with the upper and middle catchments reflecting the minimum intensity of erosivity.

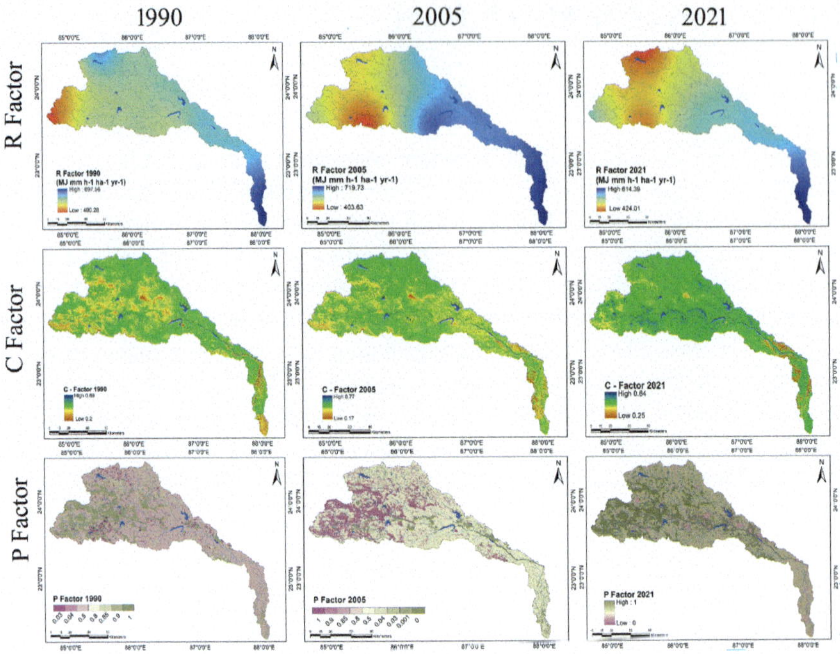

Fig. 4.1 Various factors of the RUSLE model (1) Rainfall erosivity factor (R) of 1990, 2005, and 2021 (2) Cover management factor (C) of 1990, 2005, and 2021 and (3) Support practice factor (P) of 1990, 2005 and 2021 of DRB

4.3.2.2 Soil Erodibility Factor (K-Factor)

The soil erodibility factor (K factor) is the significant variable for the RUSLE soil erosion model. K factor indicates the vulnerability of soil loss for a particular type of soil. The *K* factor reflects the soil resistance to both soil detachment and transportation [49]. Soil erodibility signifies the capacity of soil detachment, transportability of the sediment, and rate of runoff given in a particular rainfall input as a measure of standard condition. Generally, the K factor ranges from 0 to 1 (Table 4.1). The soil erodibility depends on the permeability, porosity, grain size, cohesiveness, antecedent moisture content, drainage potential, structural integrity, and organic matter of the soil. It is not possible to collect soil samples from large watersheds to estimate the soil erodibility factor. So, the erodibility maps prepared by FAO DSMW data are used to fulfill the requirements of this model [60]. Various scholars in India have calculated the soil erodibility factor using the following Eq. (4.3) [20, 32].

$$K = f_{csand} \times f_{cl-si} \times f_{orgc} \times f_{hisand} \tag{4.3}$$

Table 4.1 The K factor values for each soil groups in the Damodar River basin (DRB)

FAO soil	Particle size (%)			SOM	K factor (t ha h ha^{-1} MJ^{-1} mm^{-1})
	Sand	Silt	Clay		
Chromic Cambisols (BC)	40.1	21.5	38.4	1.44	0.09
Eutric Cambisols (BE)	36.4	37.2	26.4	1.07	0.13
Eutric Gleysols (GE)	42.8	20.4	36.8	1.3	0.09
Lithosols (I)	58.9	16.2	24.9	0.97	0.11
Eutric Fluvisols (JE)	70.8	12.8	16.5	1.15	0.10
Ferric Luvisols (LF)	74.6	9.6	15.9	0.39	0.10
Orthic Luvisols (LO)	76	9.9	14.1	0.41	0.10
Dystric Nitosols (ND)	38.9	17.6	43.6	1.57	0.08
Eutric Nitosols (NE)	68.4	10.5	21.2	0.6	0.10

where, f_{csand} act as a factor, it lowers the K indicator in soils with high coarse-sand content but elevates the K indicator for soils with little sand; f_{cl-si} gives low soil erodibility factors for soils with high clay-to-silt ratios; f_{orgc} reduces K values in soils with high organic carbon content, f_{hisand} lowers K values for soils with extremely high sand content.

$$f_{csand} = \left[0.23 + 0.3.exp\left\{-0.256m_s\left(1 - \frac{m_{silt}}{100}\right)\right\}\right] \tag{4.3a}$$

$$f_{cl-si} = \left(\frac{m_{silt}}{m_c + m_{silt}}\right)^{0.6} \tag{4.3b}$$

$$f_{orgc} = \left(1 - \frac{0.25orgC}{orgC + exp[3.72 - 2.95orgC]}\right) \tag{4.3c}$$

$$f_{hisand} = \left(1 - \frac{0.7\left(1 - \frac{m_s}{100}\right)}{\left(1 - \frac{m_s}{100}\right) + exp[-5.5 + 22.9]\left(1 - \frac{m_s}{100}\right)}\right) \tag{4.3d}$$

where,

m_s Sand fraction content (0.05 – 2.00 mm diameter) [%];
m_{silt} Silt fraction content (0.002 – 0.05 mm diameter) [%];
m_c Clay fraction content (< 0.002 mm diameter) [%];
$orgC$ the organic carbon (SOC) content [%].

In this study the Damodar River basin (DRB) consists of 10 different soil types with varying soil characteristics and the soil erodibility value (K factor) was assigned using Eq. (4.3) based on soil texture permeability and antecedent moisture content as shown in Fig. 4.2 with K values ranging from 0.076 to 0.13 (Table 4.1). A value

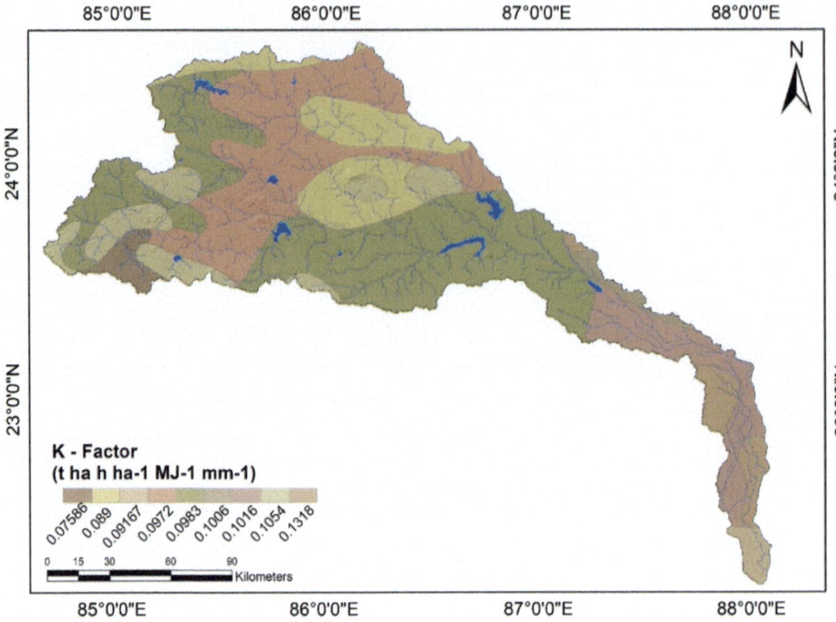

Fig. 4.2 The soil erodibility (K) factor of DRB

close to 0 indicates a lower probability of soil erosion and a value close to 1 indicates a higher probability of soil erosion.

4.3.2.3 Slope Length and Steepness Factor (LS-Factor)

The slope length and slope steepness factor are significant for sediment load and soil erosion of a particular region. This factor depends on two variables 1. Slope length (L) and 2. Slope Steepness (S) where the slope length increases, soil erosion also increases and vice versa. The geomorphology of the study area contributes significantly to soil erosion through slope length (L) and slope steepness (S) [15]. In the present study, the LS factor is calculated from DEM (spatial resolution: 30 m) following the method of [41] in ArcGIS 10.3 software [5, 17]. The slope length factor is generated from DEM. The slope length (LS) factor of the Damodar catchment varies from 0.03 to 29.42. Maximum LS factor values are situated in the upper and middle catchment of DRB (Fig. 4.3).

$$LS = (\textit{flow accumulation} \times \textit{cell size}/22.3)^{0.4} \times (\sin \textit{slope}/0.0896)^{1.3} \quad (4.4)$$

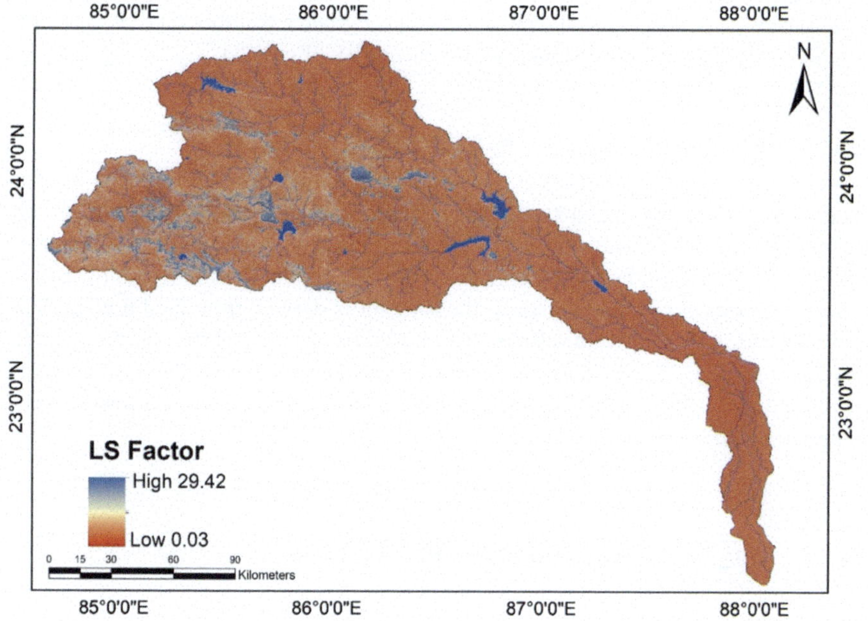

Fig. 4.3 The slope length and steepness (LS) factor of DRB

4.3.2.4 Cover Management Factor (C-Factor)

The cover management factor depends on vegetation coverage and land use land cover in a particular area. This C factor significantly influences soil erosion along with various slope lengths and steepness (Benkobi et al. 1994). The C factor is defined as the ratio of soil loss from agricultural lands to the equivalent soil loss from continuously fallow and clean-tilled lands [60] and was calculated for the Damodar River basin (DRB) using the method proposed by [55] from Eq. (4.5). The normalized difference vegetation index (NDVI) was derived from LANDSAT TM5 and LANDSAT 8 OLI data for the Damodar catchment.

$$NDVI = \frac{NIR - RED}{NIR + RED} \tag{4.5}$$

$$C = exp\left[-\alpha \frac{NDVI}{(\beta - NDVI)}\right] \tag{4.6}$$

where, the best representative values of α, and β are 2, and 1, respectively, which give a reasonable result [55]. In this study, the C factors for the Damodar catchment were calculated for 1990, 2005, and 2021, with values ranging from 0 to 1. The C factor in 1990 ranged from 0.2 to 0.69, while in 2005, it varied from 0.17 to 0.77, and in 2021 it ranged from 0.25 to 0.64. The NDVI and C factors exhibit an inverse relationship, where a higher C value indicates a greater likelihood of soil erosion, while a lower C value corresponds to a reduced risk of erosion.

4.3.2.5 Conservation Support Practice Factor (P-Factor)

The conservation practice factor (P) is called the soil-loss ratio concerning definite conservation practice [49]. Various conservation practices help to reduce the probability of soil erosion by influencing the drainage pattern, concentration of surface runoff, and runoff rate. In this study, the P-factor is derived from land use and land cover (LULC). P-factor values for each LULC class are derived from multiple reference sources [22, 25, 39, 54, 60, 64]. These P's values range from 0 to 1 (Table 4.2), with higher *P* values indicating greater soil erosion vulnerability and lower values reflecting reduced vulnerability.

4.3.3 Soil Erosion of Damodar River Basin

The distribution of soil erosion vulnerability in the Damodar River Basin (DRB) was assessed using the RUSLE empirical model for the years 1990, 2005, and 2021 (Fig. 4.4). The estimated average potential soil loss in the DRB ranged from 0.04 to 279.08 t ha^{-1} yr^{-1} in 1990, 0.03 to 263.97 t ha^{-1} yr^{-1} in 2005, and 0 to 178.78 t ha^{-1} yr^{-1} in 2021 (Fig. 4.4). Although peak soil erosion levels fluctuated from 1990 to 2021, the average erosion rate steadily increased over this period. The southwestern and middle catchments of the DRB consistently contributed significant amounts of soil erosion to the river and its dams throughout the period of analysis. The highest levels of soil erosion were observed in the upper and middle catchments of the Damodar River, as well as in the middle catchment of the Barakar Basin, which contains numerous mining areas. The Hazaribag, Kodarma Plateau, and Deoghar Uplands exhibited the greatest soil erosion potential. In contrast, the lower catchment with its lower elevation experienced minimal soil erosion.

In this study, soil erosion severity was categorized into five distinct levels—Very High, High, Moderate, Low, and Very Low—across the years 1990, 2005, and 2021 (Fig. 4.5) to capture the spatial distribution of the variable rate of erosion. The findings reveal that in 1990, over 74% of the area was classified under very low erosion potential, which was reduced to 72% in 2005 and further to 67% in 2021, reflecting an increasing trend of erosion risk and widespread the influence of human-induced land modifications in the DRB. In contrast, the percentage of land classified as having very high to high erosion potential rose from 0.9% in 1990 to 1% in 2005, and further to 1.5% in 2021. This increase in high-risk erosion zones over time highlights the intensification of erosion vulnerability from 1990 to 2021.

During a recent field visit (17.02.2024) to the Tenughat Dam catchment area in Jharkhand, we identified several regions highly vulnerable to soil erosion. This vulnerability is attributed to the combination of high-intensity rainfall and fragile

Table 4.2 P factor values for different land use classes in the Damodar River basin (DRB) for the years 1990, 2005, and 2021

Year	LULC	P factor	References
1990	Agricultural fallow	0.85	[39]
	Built-up Area	0.001	[25]
	Sand	0.85	[22]
	Open forest	0.8	[54]
	Bare land	1	[54]
	Dense forest	1	[54]
	Agricultural land	0.85	[64]
	Water Body	0	[60]
2005	Barren land	1	[54]
	Built-up Area	0.001	[25]
	Cropland	0.85	[64]
	Deciduous forest	1	[54]
	Fallow land	0.85	[39]
	Grassland	0.04	[39]
	Mixed forest	0.8	[54]
	Plantation	1	[54]
	Shrubland	1	[54]
	Wasteland	0.9	[54]
	Water Body	0	[60]
2021	Agricultural land	0.85	[64]
	Bare land	1	[54]
	Built-up Area	0.001	[25]
	Dense forest	0.9	[54]
	Flooded vegetation	1	[54]
	Open forest	0.8	[54]
	Shrubland	1	[54]
	Water body	0	[60]

rock formations. Ranchi experiences an annual rainfall that typically ranges between 1200 and 1600 mm. The topography is predominantly undulating, with terraced slopes and fragile mountain soils further exacerbating the erosion process (Fig. 4.6). These conditions have created multiple soil erosion hotspots (SEH), where the interplay of climatic and geomorphological factors accelerates land degradation. The fragile nature of the soil, coupled with steep slopes and heavy rainfall, makes the region highly susceptible to sediment loss. These findings emphasize the critical need for targeted soil conservation measures and land management strategies to mitigate erosion and protect the environment in this region.

Table 4.3 Roughness coefficient (n) of various land use land cover [4, 26, 63]

Year	LULC	Manning roughness coefficient (n)
1990 and 2005	Deciduous broadleaf forest	0.76
	Croplands	2.62
	Built-up area	5.14
	Mixed forest	0.76
	Shrub land	0.75
	Barren land	3.88
	Fallow land	3.88
	Wasteland	3.88
	Water body	4.91
	Plantation	0.75
	Grassland	2.13
2021	Water	4.91
	Trees	0.76
	Grass	2.13
	Flooded vegetation	0.76
	Crops	2.62
	Shrubs	0.75
	Built-up area	5.14
	Bare land	3.88

4.4 Estimation of Sediment Delivery Ratio (SDR)

4.4.1 SDR Model

The sediment delivery ratio (SDR) is one of the most significant factors that enormously affect the sediment yield (SY) to the channel. In large watersheds, the maximum sediments are deposited in various spatial areas before reaching the outlets [18]. The Damodar catchment is regarded as a large watershed, and its sediment yield dynamics are highly complex due to the expansive upper catchment area and the variability in monsoonal rainfall patterns. Calculation of SDR on the large watershed scale remains a challenge till now. The sediment distributed delivery (SEDD) model estimates the probability of eroded particles in a particular location being delivered to the nearest channel or outlets [21]. The SDR model is proven to be efficient for calculating both spatial and temporal variations in sediment transport within the Damodar catchment. The SEDD model proposed by FERRO and MINACAPILLI [21] determined SDR in every single pixel in the catchment areas. The SDR model is used to assess spatial and temporal variations in sediment transport [21, 28] following Eq. (4.7).

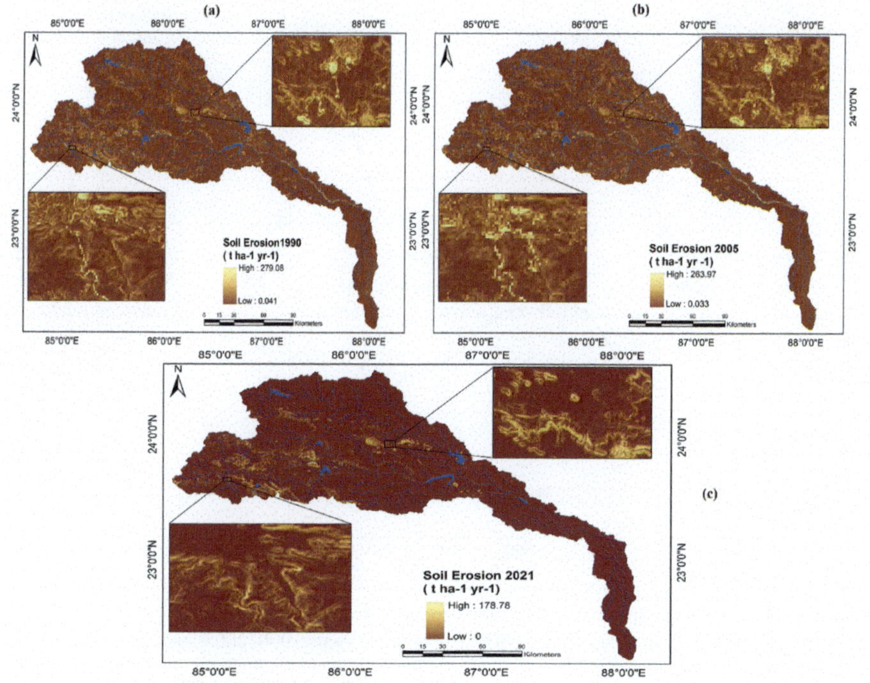

Fig. 4.4 Spatial distribution of annual soil loss rate (t ha^{-1} yr^{-1}) in the Damodar River basin in the years of a. 1990, b. 2005, and c. 2021

$$SDR_i = \exp(-\beta t_i) \tag{4.7}$$

where,

SDR_i	Sedimentary Delivery Ratio
β	Basin – specific parameter
t_i	Travel time (hr) for cell i.

The travel time of overland flow from a pixel to the nearest main channel or stream channel along the flow path was computed as followed by Eq. (4.8).

$$t_i = \sum_{i=1}^{m} \frac{l_i}{V_i} \tag{4.8}$$

where,

l_i	Flow length (m) of the cell i
V_i	Flow velocity for the cell (ms^{-1}).

The flow length (m) was calculated using the D8 pour point algorithm implemented in the TauDEM 5.3.7 add-in tool for ArcGIS 10.3 software [52]. A threshold

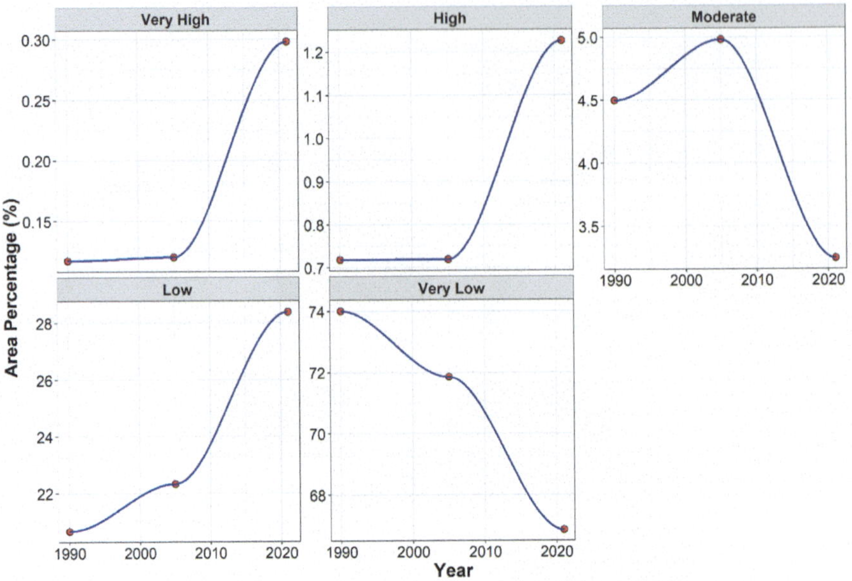

Fig. 4.5 The changing pattern of soil erosion severity classes (very high, high, moderate, low, and very low) in the Damodar River basin (DRB) between the years 1990, 2005, and 2021

Fig. 4.6 The fragile mountain soils in the Patratu Valley region (SEH) of Jharkhand exhibit a high vulnerability to soil erosion (Photo credit: Sk Asraful Alam)

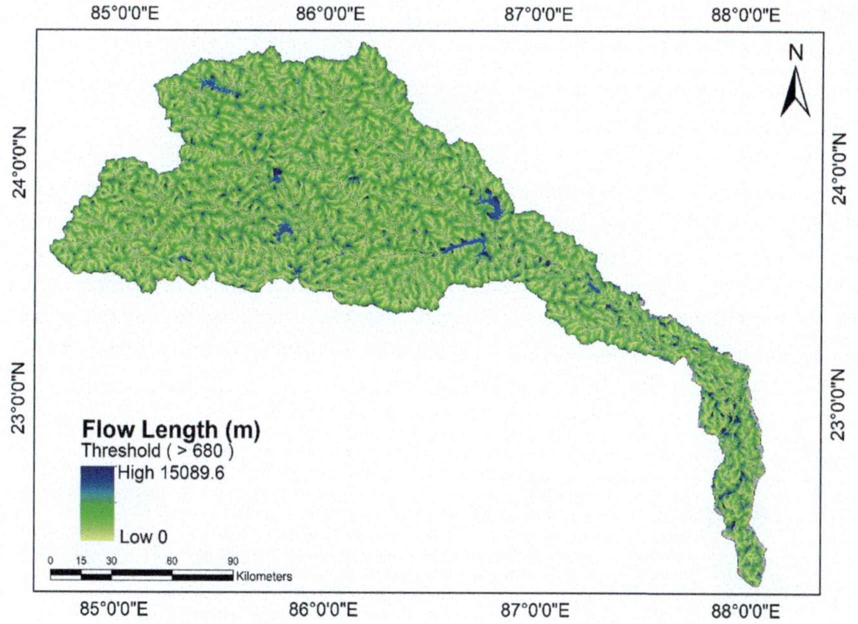

Fig. 4.7 Flow length map of the Damodar River basin (DRB)

value of greater than 680 was employed for calculating the grid-based flow length (Fig. 4.7). Flow length represents the distance water travels along the surface from a given point to the nearest drainage outlet or watershed boundary. Additionally, the flow velocity (ms^{-1}) was determined to assess the SDR in the Damodar catchment, following the methodology outlined by [13].

$$V_i = a_i S_i^b \tag{4.9}$$

where,

a_i Surface roughness coefficient (n)
S_i Slope gradient
b It is a numerical constant equal to 0.5 [13, 21].

The coefficient value for a_i (n) is derived from land use classifications, which are assigned based on the land use maps [26].

4.4.2 *SDR of Damodar Catchment*

The sediment distributed delivery (SEDD) model has been employed for the estimation of sediment delivery ratio (SDR) in the Damodar catchment for the years 1990, 2005, and 2021 (Fig. 4.8). The sediment delivery ratio (SDR) continuously decreased over time for the year of 1990–2021. The upper and middle catchments of the Damodar basin show the maximum SDR whereas lower catchment areas show the minimum SDR. Slope and topography greatly influence the sediment delivery of a catchment. A higher slope results in the higher LS factor which indicates the maximum sediment delivery to the channel. Therefore, the upper and middle catchments of the Damodar basin display the maximum sediment delivery, and the lower catchment displays minimum sediment delivery.

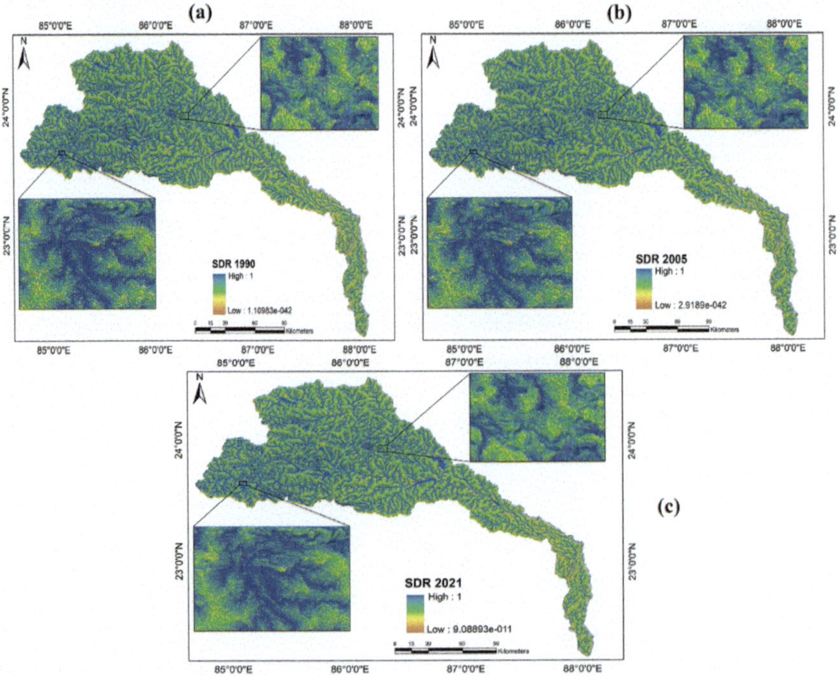

Fig. 4.8 Sediment delivery ratio (SDR) in the Damodar River basin for the years 1990, 2005, and 2021

4.5 Relationship Between SDR and SY

Sediment yield (SY) and sediment delivery ratio (SDR) of the Damodar River basin (DRB) are calculated for the years 1990, 2005, and 2021. Damodar catchment has been classified into 23 sub-basins. The 1500 random points have been generated to collect the sample values of SY and SDR in the Damodar catchment to understand the relationship between these two variables of every single sub-basin. The relationship between the two variables has significantly increased over time (1990–2021) which indicates the maximization of sediment input to the channels and dams. Sub-basins no. SB7 indicates the highest correlation ($R^2 = 0.43$) where the SB1, SB4, SB9, SB11, SB14, and SB16 have the R^2 value of 0.22, 0.18, 0.26, 0.19, 0.21 and 0.21 respectively, for the year of 1990 (Fig. 4.9). In the year of 2005, the SB7 has the highest correlation ($R^2 = 0.54$) where the SB1, SB4, SB9 and SB16 have the R^2 value of 0.20, 0.19, 0.18 and 0.34 respectively (Fig. 4.10). In the year of 2021, the SB7 has the highest correlation ($R^2 = 0.50$) where the SB1, SB3, SB6, SB8, SB13 and SB16 have the R^2 value of 0.20, 0.28, 0.24, 0.22, 0.29 and 0.36 respectively (Fig. 4.11). The upper and middle catchment sub-basins of the Damodar basin exhibit a significant increasing trend in the correlation between sediment delivery ratio (SDR) and sediment yield (SY) for the period from 1990 to 2021. Sub-basins SB1, SB4, and SB14 demonstrate strong correlations throughout this timeframe, and all of these sub-basins are associated with the Maithon dam.

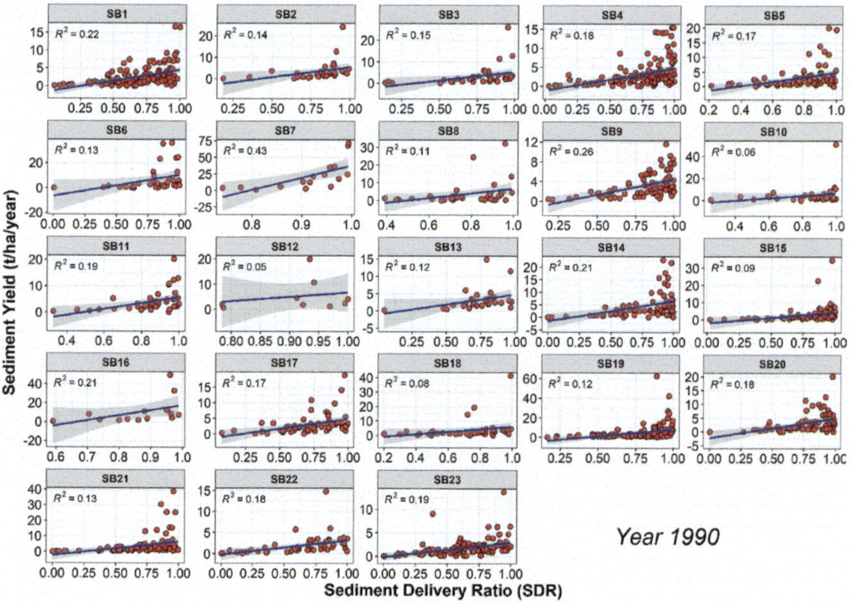

Fig. 4.9 Sub-basin-wise correlation between SDR and SY in the year 1990

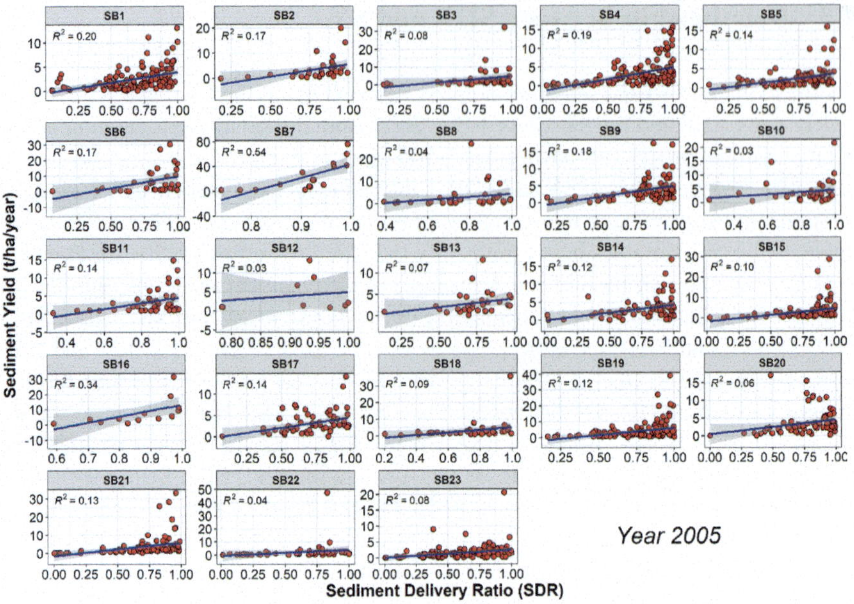

Fig. 4.10 Sub-basin-wise correlation between SDR and SY in the year 2005

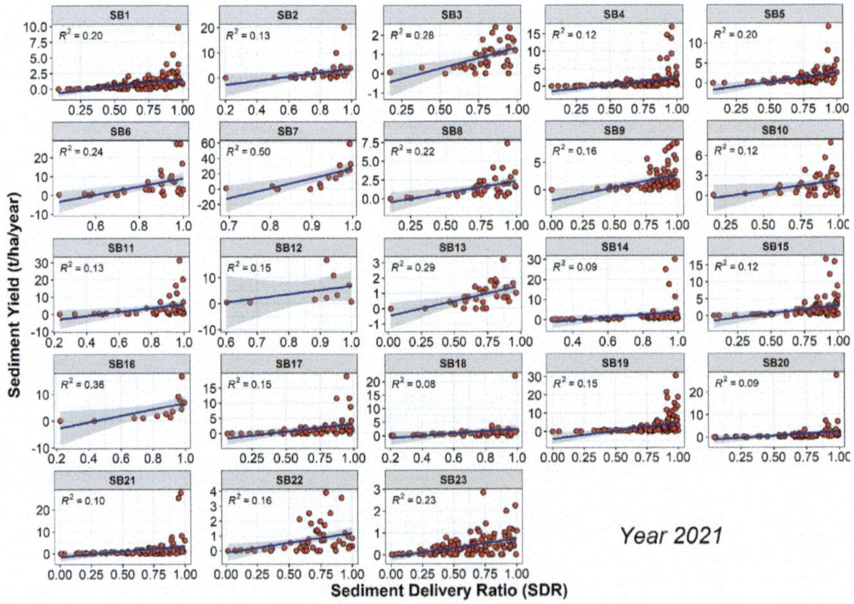

Fig. 4.11 Sub-basin-wise correlation between SDR and SY in the year 2021

4.6 Relationship Between SE and SY

Sediment yield (SY) and soil erosion (SE) were successfully calculated for the Damodar River Basin in the years 1990, 2005, and 2021. The sediment yield to the channels or outlets is influenced by the intensity of soil erosion and surface runoff, leading to sedimentation in the channel. To understand the dynamics of sediment transport, a relationship between SY and SE was established for all sub-basins within the Damodar basin. This correlation varies among sub-basins due to factors such as drainage area, topographical conditions, drainage density, and land cover. In this study, we developed the correlation between SE and SY across 23 sub-basins of the Damodar River Basin for the years 1990, 2005, and 2021. The relationship between these two variables has significantly strengthened over time, indicating an increase in sediment input to both the channels and dams. In 1990, all sub-basins displayed a strong correlation ($R^2 > 0.95$), except for SB22 and SB23 (Fig. 4.12). In 2005, all sub-basins exhibited an even higher correlation ($R^2 > 0.97$), except for SB23 (Fig. 4.13). In 2021, the correlations reached their highest levels ($R^2 > 0.98$) for all sub-basins, again excluding SB23 (Fig. 4.14). The sub-basins of the upper and middle catchments of the Damodar River basin show a significant increasing trend in the correlation between soil erosion (SE) and sediment yield (SY) from 1990 to 2021. While SB22 and SB23 exhibited lower correlations compared to the other sub-basins in 1990, these correlations have markedly improved in 2005 and 2021.

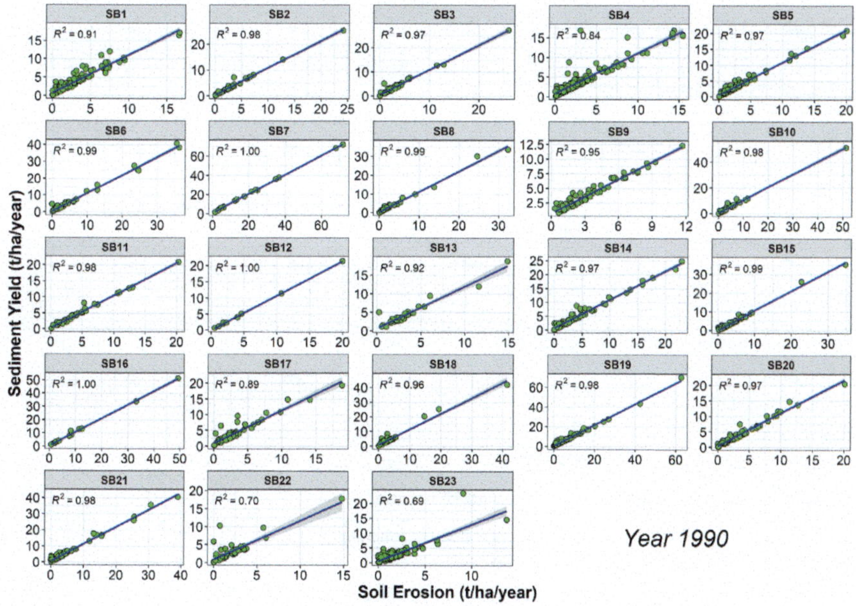

Fig. 4.12 Sub-basin-wise relationship between SE and SY in the year 1990

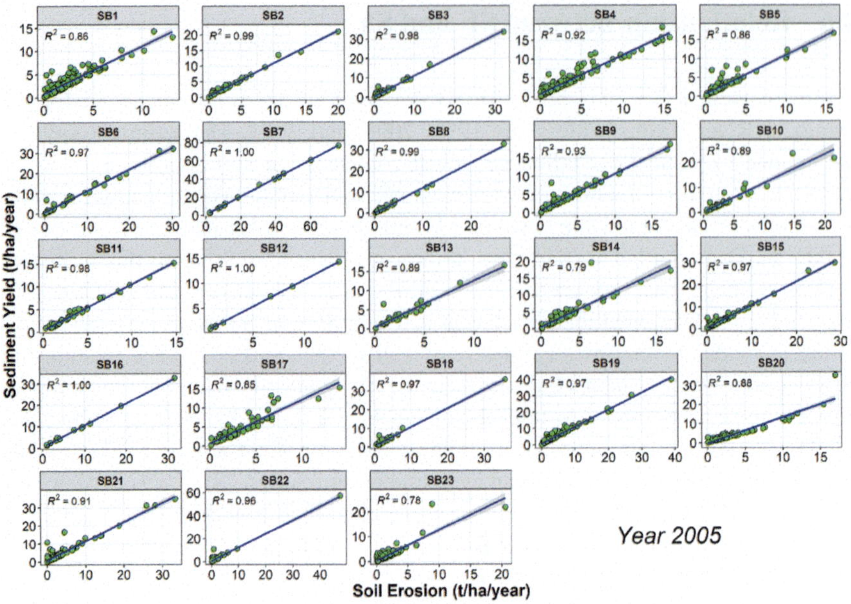

Fig. 4.13 Sub-basin-wise relationship between SE and SY in the year 2005

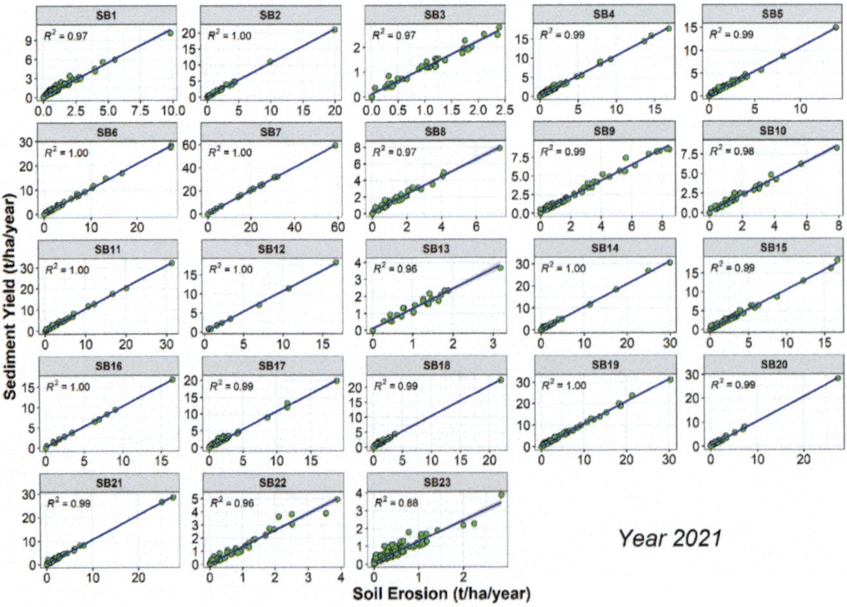

Fig. 4.14 Sub-basin-wise relationship between SE and SY in the year 2021

4.7 Identification of Soil Erosion Hotspot (SEH) in DRB

The identification of hotspots is nothing but the spatial analysis technique that identifies the statistically significant ($p < 0.05$) clustering pattern of high z-values (Hot spots) and low z-values (Cold spots). In the present study, we apply the Getis-Ord Gi* statistics for the identification of soil erosion hotspots (SEH) in the Damodar River basin (DRB). The SEH is significant for maximum sediment input to the channels and siltation input to the dams. Therefore, the identification of soil erosion hotspots (SEH) in DRB might be helpful for the management of sedimentation and siltation in various dams like Maithon, Panchet, Konar, Tenughat, and Tilaya.

4.7.1 Methods

In the present study, the Getis-Ord Gi* statistics [24] were applied to the Damodar catchment to identify the soil erosion hotspot (SEH). These methods find the spatial association among neighboring sample points and identify the statistically significant ($p < 0.05$) spatial clusters of high and low point values as hotspots and coldspots, respectively [24]. The outputs of Getis-Ord Gi* statistics identify feature classes through Z-score, *P*-value, and Confidence Level (CL). The features with a high Z-score and small *P*-value indicate statistically significant hotspots and features with a low Z-score value and small *P*-value indicate statistically significant coldspots. The Getis-Ord Gi* method is followed by the Eq. (4.10).

$$G_i^* = \frac{\sum_{j=1}^{n} w_{i,j}x_j - \overline{X}\sum_{j=1}^{n} w_{i,j}}{S\sqrt{\frac{\left[n\sum_{i,j}^{n} w_{i,j}^2 - \left(\sum_{i,j}^{n} w_{i,j}\right)^2\right]}{n-1}}} \tag{4.10}$$

where, the G_i^* is a z-score of patch i. x_j is the attribute value for patch j; w_{ij} is the spatial weight between patch i and patch j, if the distance from a neighbor j to the feature i is within the distance, $w_{ij} = 1$; otherwise $w_{ij} = 0$; n is the total number of grid cells and

$$\overline{X} = \frac{\sum_{j=1}^{n} x_j}{n} \tag{4.11}$$

$$S = \sqrt{\frac{\sum_{j=1}^{n} x_j^2}{n} - \left(\overline{X}\right)^2} \tag{4.12}$$

Identifying and mapping soil erosion hotspots (SEH) in the Damodar catchment can offer valuable insights for targeted policymaking aimed at managing substantial sedimentation in reservoirs and dams. [14] have indicated that utilizing hotspot analysis through a random approach could increase conservation benefits by as much as 25%. Therefore, identifying soil erosion hotspots is essential for effectively managing sedimentation and protecting fluvial ecosystems.

4.7.2 SEH in Damodar River Basin

The soil erosion hotspot (SEH) has been determined by Eq. (4.10) in the Damodar River basin (DRB) for the years 1990, 2005, and 2021 (Fig. 4.15). Hotspots are mainly concentrated in the Hazaribag, Kodarma, and Deoghar upland and Parasnath hill regions in the upper and middle catchment in DRB. The Soil erosion hotspot (SEH) areas overlap with the relatively very high, high, and relatively high soil erosion zones. Statistically significant (P-value < 0.01) hotspots with a 99% confidence interval (CI) cover 9% of the study area in the year 1990, where 7% and 10.2% in the years 2005 and 2021 respectively. The areal distribution of SEH in the Damodar catchment for the years 1990, 2005, and 2021 indicates a significantly increasing trend. The area distribution of soil erosion hotspots (SEH) patches significantly increased in the year 2021. Whereas it was confined to a few small patches in 2005.

The soil erosion severity pixels have been clustered by Getis-Ord Gi^* statistics which indicates the soil erosion hotspot (SEH) patches in the Damodar River basin (DRB). The Getis-Ord Gi^* statistics provide the standard deviation (SD), Z-score value, and P-value for the entire soil erosion pixels. The pixels showing a high SD, high Z-score, and small P-value indicate statistically significant hotspots, and features with high SD, low Z-score value and small P-value indicate statistically significant coldspots (Fig. 4.16).

The spatial distribution of Soil Erosion Hotspot (SEH) locations in the Damodar River Basin (DRB) has altered significantly between 1990 and 2021. The majority of the hotspot areas are concentrated in the upper and middle catchments of the Damodar River Basin (Fig. 4.17), particularly within the contributing areas of the Maithon, Panchet, and Tenughat dams. In comparison, the Konar and Tilaya dams include only a small portion of the hotspot areas. Soil erosion hotspot areas have expanded over time within the catchments of the Tenughat, Panchet, and Maithon dams. The Tenughat dam's catchment includes sub-basins SB8, SB11, SB12, SB15, and SB19, while the Panchet dam covers SB6, SB5, SB9, SB10, SB17, SB18, and SB20, and the Maithon dam includes SB1, SB2, SB3, SB4, and SB14. The Tenughat dam experiences a considerable increase in the soil erosion hotspot area by 1250 km^2 (Fig. 4.18).

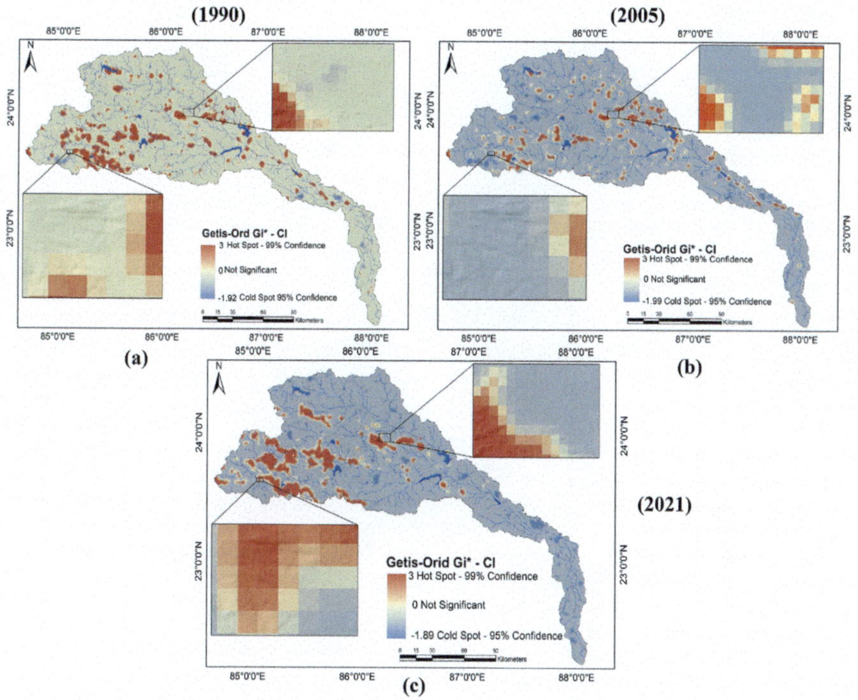

Fig. 4.15 The soil erosion hotspot (SEH) map of the Damodar River Basin (DRB) for the years 1990, 2005, and 2021

4.8 The Effect of Soil Erosion Hotspots on Dams

The Damodar River Basin (DRB) was divided into 23 sub-basins using threshold elevation values in ArcGIS. Soil erosion (SE), Sediment Delivery Ratio (SDR), and Sediment Yield (SY) were assessed for the years 1990, 2005, and 2021 (Fig. 4.19). Sub-basins SB6, SB7, SB11, SB12, SB16, and SB19 exhibited above-average soil erosion rates for the entire DRB, and correspondingly, these sub-basins also had higher-than-average SDR values across all three years. Notably, the highest soil erosion rates were observed in SB7, SB12, SB16, and SB19, which also recorded the highest SDR values: SB7 (0.86), SB12 (0.82), SB16 (0.82), and SB19 (0.78) over the study period. This indicates that these sub-basins not only experienced higher erosion but also delivered more sediment to downstream areas, highlighting them as critical zones for soil conservation efforts in the basin.

The effect of soil erosion hotspots on dams is significant as these areas contribute to increased sediment yield and sediment delivery, which can negatively impact reservoir storage capacity and dam operations. In the Damodar River Basin (DRB), sub-basins with high soil erosion rates, such as SB7, SB12, SB16, and SB19, also exhibit elevated sediment delivery ratios (SDR), meaning that these regions are major

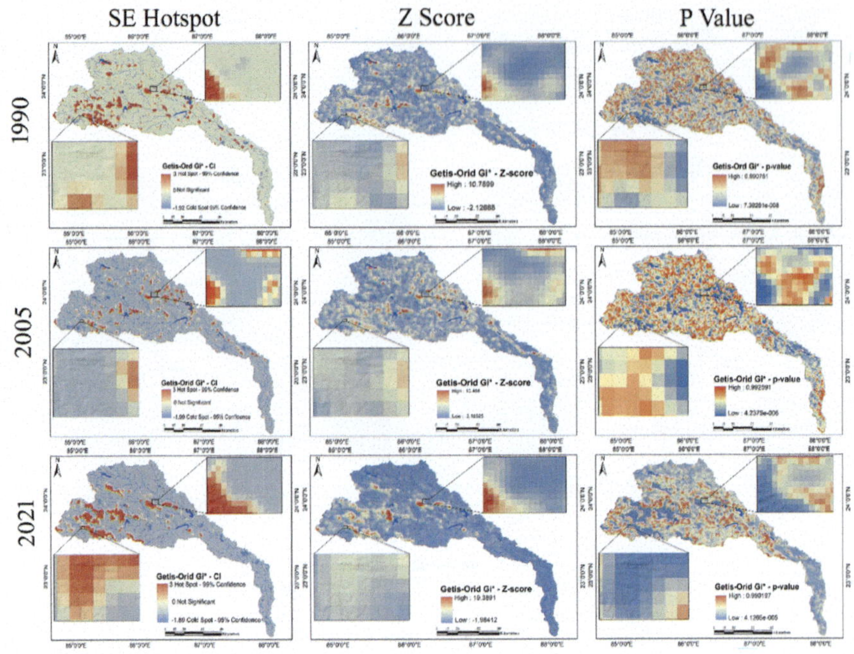

Fig. 4.16 The soil erosion hotspot (SEH) map in DRB at various statistical scales 1. SEH map, 2. Z-score map, and 3. *P*-value map of 1990, 2005, and 2021

contributors of sediment to the downstream reservoirs. Over the years 1990, 2005, and 2021, the erosion and sediment delivery from these hotspots have increased, potentially leading to accelerated sedimentation in the dams. This can reduce the effectiveness of the dams for water storage, flood control, and power generation, highlighting the need for targeted soil erosion management and conservation strategies in these critical areas to mitigate the long-term impacts on dam infrastructure and ecosystem services.

4.9 Major Findings

The major findings of this chapter provide crucial insights for developing future management plans for the Damodar River Basin (DRB). The key takeaways from this chapter are:

- The average annual rates of soil erosion (SE), sediment delivery ratio (SDR), and sediment yield (SY) across the entire Damodar catchment from 1990 to 2021 were 39, 0.75, and 32.12 t ha^{-1} yr^{-1} respectively.

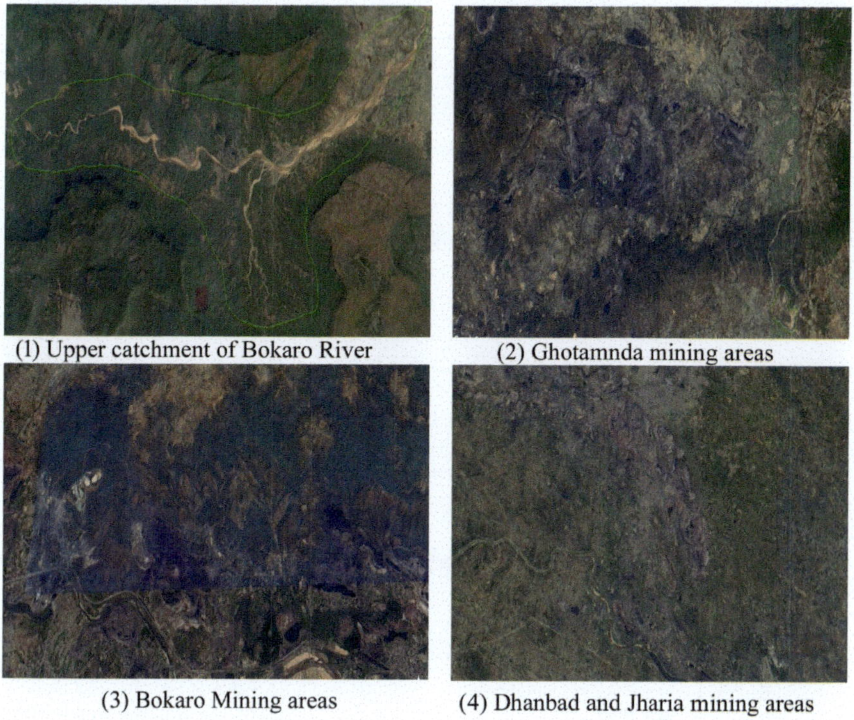

Fig. 4.17 Google Earth image shows soil erosion hotspot in Damodar River basin: (1) The upper catchment of Bokaro River, (2) Ghotamnda mining areas, (3) Bokaro mining areas, and (4) Dhanbad and Jharia mining areas

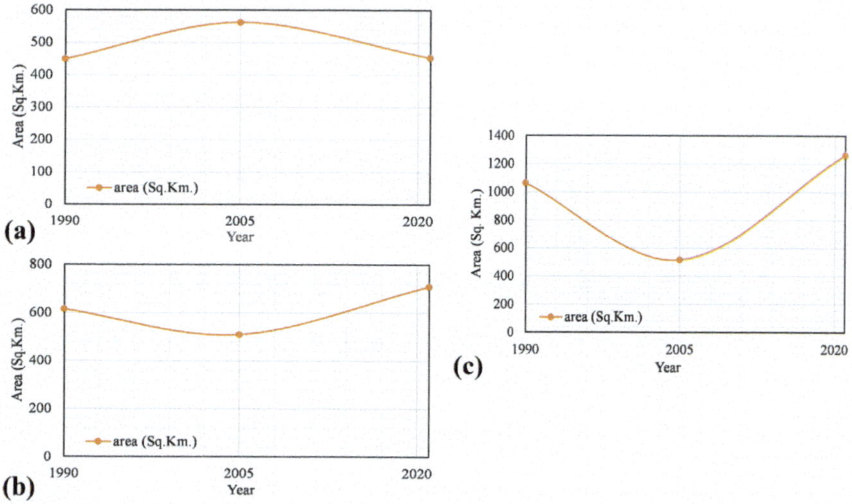

Fig. 4.18 The map shows the areal distribution of SEH for the years 1990, 2005, and 2021 in the dam of (a) Maithon, (b) Panchet, and (c) Tenughat

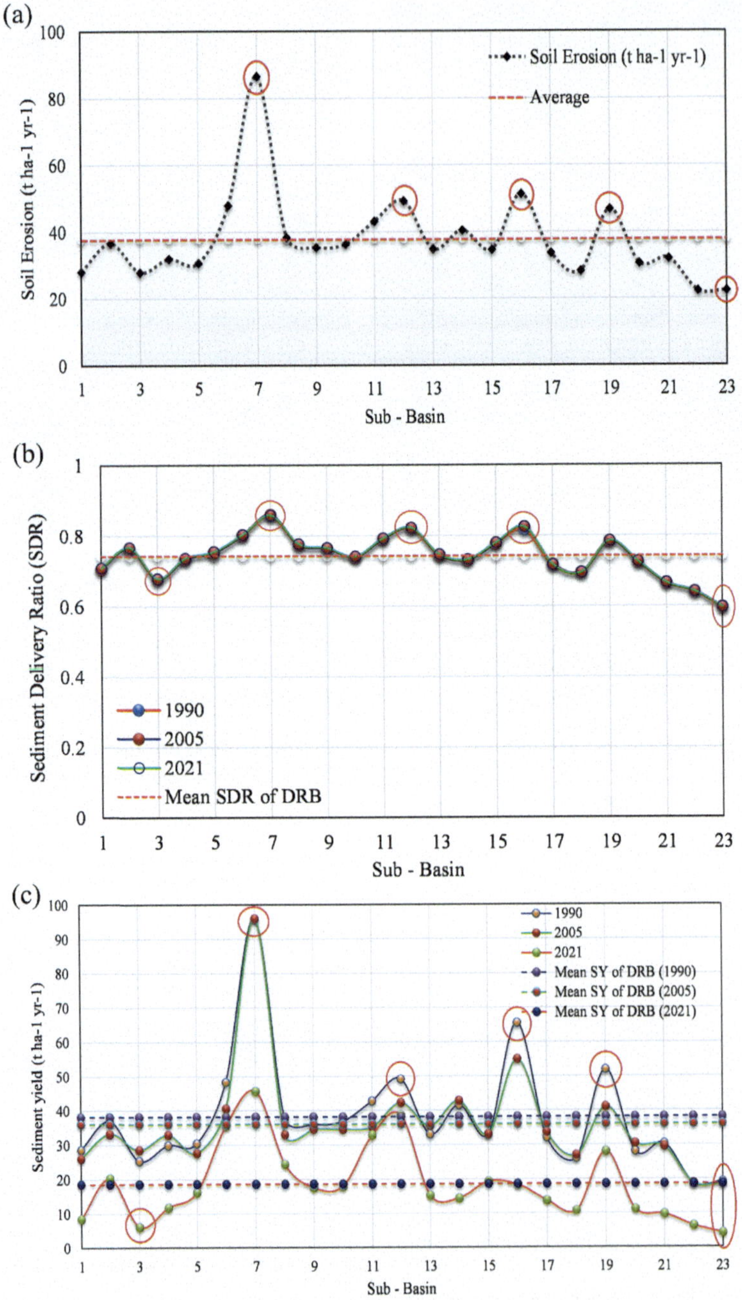

Fig. 4.19 The graph showing the (a) average soil erosion (SE) rate of all sub-basins in Damodar catchment, (b) average sediment delivery ratio (SDR) of all sub-basins in DRB for the year of 1990, 2005 and 2021, and (c) average sediment yield (SY) of all sub-basins in DRB for the year of 1990, 2005 and 2021

- Sub-basins SB6, SB7, SB11, SB12, SB16, and SB19 consistently demonstrated the highest mean values for SE, SY, and SDR during 1990, 2005, and 2021 across the catchment.
- The upper and middle catchments exhibited the highest levels of soil erosion, sediment yield, and SDR, largely due to undulating terrain, mining activity, steep slopes, deforestation, and agricultural practices.
- Sub-basin SB7 showed the strongest positive correlation between SDR and SY, and between SE and SY, over the years, driven by the intense mining activities in the Bokaro region, where severe soil erosion was observed.
- Among the five dams in the Damodar catchment (Maithon, Panchet, Tenughat, Konar, and Tilaya), the contributing areas of Tenughat and Panchet dams were identified as having the highest probability of severe soil erosion.
- Soil erosion hotspot (SEH) areas were primarily concentrated in the Hazaribag, Kodarma, and Deoghar uplands, as well as the Parasnath hill region within the upper and middle catchments of the DRB.

4.10 Conclusion

This chapter focuses on estimating the soil erosion rate (SE), sediment delivery ratio (SDR), sediment yield (SY), and soil erosion hotspots (SEH) within the Damodar River Basin (DRB) for the years 1990, 2005, and 2021. The study's principal aims are to evaluate the relationship between soil erosion and sediment output across 23 subbasins, as well as to identify and spatially map soil erosion hotspot (SEH) patches within catchment areas. The average gross soil erosion rate, determined using the RUSLE model, was 46.53 t ha^{-1} yr^{-1} in 1990, 45.32 t ha^{-1} yr^{-1} in 2005, and 25.20 t ha^{-1} yr^{-1} in 2021. The SDR, estimated using the SEDD model, was 0.7426 in 1990, 0.7425 in 2005, and 0.7441 in 2021. The hotspot patches reveal significant zones of severe erosion and are primarily concentrated in the Hazaribag Kodarma and Deoghar uplands and the Parasnath hill regions in the upper and middle catchments of the DRB.

The detailed analysis of SE, SDR, and SEH highlights several sub-basins with a high probability of contributing sediment to the main channel as well as reservoirs. The Maithon, Panchet, and Tenughat dams recorded SEH areas of 450, 615.90, and 1065.01 km^2 in 1990, increasing to 453.97, 709.93, and 1264.94 km^2 in 2021, respectively. The expansion of SEH areas between 1990 and 2021 indicates a significant increasing trend in sediment production in the upper catchment, which increases the probability of sediment load to the main channel and threatens the capacity of major dams in the DRB. Identifying these SEH patches is essential for understanding the severity of soil erosion and its direct impact on sediment yield and siltation in reservoirs. The findings provide valuable insights for developing effective siltation management strategies in the Damodar catchment and offer crucial guidance for policymakers in managing SEH areas efficiently.

References

1. Achu AL, Thomas J (2023) Soil erosion and sediment yield modeling in a tropical mountain watershed of the southern Western Ghats, India using RUSLE and Geospatial tools. Total Environ Res Themes 8:100072. https://doi.org/10.1016/j.totert.2023.100072
2. Arnold JG, Srinivasan R, Muttiah RS, Williams JR (1998) Large area hydrologic modeling and assessment Part I: model development. J Am Water Resour Assoc 34:73–89
3. Babu R, Dhyani BL, Nirmal K (2004) Assessment of erodibility status and refined iso-erodent map of India. Indian J Soil Conserv 32:171–177
4. Batista PVG, Silva MLN, Silva BPC et al (2017) Modelling spatially distributed soil losses and sediment yield in the upper Grande River Basin—Brazil. CATENA 157:139–150. https://doi.org/10.1016/j.catena.2017.05.025
5. Benchettouh A, Kouri L, Jebari S (2017) Spatial estimation of soil erosion risk using RUSLE/GIS techniques and practices conservation suggested for reducing soil erosion in Wadi Mina watershed (northwest, Algeria). Arab J Geosci 10:79. https://doi.org/10.1007/s12517-017-2875-6
6. Bhattacharya RK, Das CN, Das K (2020) Estimation of erosion susceptibility and sediment yield in ephemeral channel using RUSLE and SDR model: tropical Plateau Fringe region, India. Adv Sci Technol Innov 163–185. https://doi.org/10.1007/978-3-030-23243-6_10
7. Biswas SS, Pani P (2015) Estimation of soil erosion using RUSLE and GIS techniques: a case study of Barakar River basin, Jharkhand, India. Model Earth Syst Environ 1:1–13. https://doi.org/10.1007/s40808-015-0040-3
8. Borrelli P, Alewell C, Alvarez P et al (2021) Soil erosion modelling: a global review and statistical analysis. Sci Total Environ 780:146494. https://doi.org/10.1016/j.scitotenv.2021.146494
9. Boubekraoui H, Maouni Y, Ghallab A et al (2023) Spatio-temporal analysis and identification of deforestation hotspots in the Moroccan western Rif. Trees, For People 12:100388. https://doi.org/10.1016/j.tfp.2023.100388
10. Chen W, Huang Y-C, Lebar K, Bezak N (2023) A systematic review of the incorrect use of an empirical equation for the estimation of the rainfall erosivity around the globe. Earth-Science Rev 238:104339. https://doi.org/10.1016/j.earscirev.2023.104339
11. Chicas SD, Omine K, Ford JB (2016) Identifying erosion hotspots and assessing communities' perspectives on the drivers, underlying causes and impacts of soil erosion in Toledo's Rio Grande Watershed: Belize. Appl Geogr 68:57–67. https://doi.org/10.1016/j.apgeog.2015.11.010
12. Chuenchum P, Xu M, Tang W (2019) Estimation of soil erosion and sediment yield in the Lancang-Mekong River using the modified revised universal soil loss equation and GIS techniques. Water 12:135. https://doi.org/10.3390/w12010135
13. Cronshey R (1986) Urban hydrology for small watersheds. US Dept Agric Soil Conserv Serv Eng Div 164
14. Crossman ND, Bryan BA (2009) Identifying cost-effective hotspots for restoring natural capital and enhancing landscape multifunctionality. Ecol Econ 68:654–668. https://doi.org/10.1016/j.ecolecon.2008.05.003
15. da Cunha ER, Bacani VM, Panachuki E (2017) Modeling soil erosion using RUSLE and GIS in a watershed occupied by rural settlement in the Brazilian Cerrado. Nat Hazards 85:851–868. https://doi.org/10.1007/s11069-016-2607-3
16. Diwediga B, Le QB, Agodzo SK et al (2018) Modelling soil erosion response to sustainable landscape management scenarios in the Mo River Basin (Togo, West Africa). Sci Total Environ 625:1309–1320. https://doi.org/10.1016/j.scitotenv.2017.12.228
17. Dutta D, Das S, Kundu A, Taj A (2015) Soil erosion risk assessment in Sanjal watershed, Jharkhand (India) using geo-informatics, RUSLE model and TRMM data. Model Earth Syst Environ 1:37. https://doi.org/10.1007/s40808-015-0034-1

18. Ebrahimzadeh S, Motagh M, Mahboub V, Mirdar Harijani F (2018) An improved RUSLE/SDR model for the evaluation of soil erosion. Environ Earth Sci 77:454. https://doi.org/10.1007/s12665-018-7635-8
19. Erkossa T, Wudneh A, Desalegn B, Taye G (2015) Linking soil erosion to on-site financial cost: lessons from watersheds in the Blue Nile basin. Solid Earth 6:765–774. https://doi.org/10.5194/se-6-765-2015
20. Farhan Y, Nawaiseh S (2015) Spatial assessment of soil erosion risk using RUSLE and GIS techniques. Environ Earth Sci, 74(6):4649–4669. https://doi.org/10.1007/s12665-015-4430-7
21. Ferro V, Minacapilli M (1995) Sediment delivery processes at basin scale. Hydrol Sci J 40:703–717. https://doi.org/10.1080/02626669509491460
22. Galetovic JR, Toy TJ, Foster GR (1998) Guidelines for the use of the Revised Universal Soil Loss Equation (RUSLE) version 1.06 on mined lands, construction sites, and reclaimed lands. Office of surface mining and reclamation (OSM), Western Regional Coordinating Center, Denver, Colorado
23. Ganasri BP, Ramesh H (2016) Assessment of soil erosion by RUSLE model using remote sensing and GIS—a case study of Nethravathi Basin. Geosci Front 7:953–961. https://doi.org/10.1016/j.gsf.2015.10.007
24. Getis A, Ord JK (1992) The analysis of spatial association by use of distance statistics. Geogr Anal 24:189–206. https://doi.org/10.1111/j.1538-4632.1992.tb00261.x
25. Guobin F, Shulin C, Donald MK (2006) Modeling the impacts of no-till practice on soil erosion and sediment yield with RUSLE, SEDD, and ArcView GIS. Soil Tillage Res 85:38–49. https://doi.org/10.1016/j.still.2004.11.009
26. Haan CT, Barfield BJ, Hayes JC (1994) Design Hydrology and Sedimentology for small catchments
27. Hooke RL (2000) On the history of humans as geomorphic agents. Geology 28:843. https://doi.org/10.1130/0091-7613(2000)28%3c843:OTHOHA%3e2.0.CO;2
28. Jain MK, Kothyari UC (2000) Estimation of soil erosion and sediment yield using GIS. Hydrol Sci J 45:771–786. https://doi.org/10.1080/02626660009492376
29. Karan SK, Ghosh S, Samadder SR (2019) Identification of spatially distributed hotspots for soil loss and erosion potential in mining areas of Upper Damodar Basin—India. CATENA 182:104144. https://doi.org/10.1016/j.catena.2019.104144
30. Kolli MK, Opp C, Groll M (2021) Estimation of soil erosion and sediment yield concentration across the Kolleru Lake catchment using GIS. Environ Earth Sci 80:161. https://doi.org/10.1007/s12665-021-09443-7
31. Kothyari UC, Singh VP (1996) Rainfall and temperature trends in India. Hydrol Process 10:357–372. https://doi.org/10.1002/(SICI)1099-1085(199603)10:3%3c357::AID-HYP305%3e3.0.CO;2-Y
32. Kumar S, Kushwaha SPS (2013) Modelling soil erosion risk based on RUSLE-3D using GIS in a Shivalik sub-watershed. J Earth Syst Sci, 122(2):389–398. https://doi.org/10.1007/s12040-013-0276-0
33. Kumar S, Mishra A (2015) Critical erosion area identification based on hydrological response unit level for effective sedimentation control in a River Basin. Water Resour Manag 29:1749–1765. https://doi.org/10.1007/s11269-014-0909-3
34. Kumar S, Parida BR (2021) Hydroponic farming hotspot analysis using the Getis-Ord Gi* statistic and high-resolution satellite data of Majuli Island, India. Remote Sens Lett 12:408–418. https://doi.org/10.1080/2150704X.2021.1895446
35. Li Y, Zhang L, Yan J et al (2017) Mapping the hotspots and coldspots of ecosystem services in conservation priority setting. J Geogr Sci 27:681–696. https://doi.org/10.1007/s11442-017-1400-x
36. Lim KJ, Sagong M, Engel BA et al (2005) GIS-based sediment assessment tool. CATENA 64:61–80. https://doi.org/10.1016/j.catena.2005.06.013
37. Luo M, Jia X, Zhao Y et al (2024) Ecological vulnerability assessment and its driving force based on ecological zoning in the Loess Plateau, China. Ecol Indic 159:111658. https://doi.org/10.1016/j.ecolind.2024.111658

38. Maner SB (1958) Factors affecting sediment delivery rates in the red hills physiographic area. Eos Trans Am Geophys Union 39:669–675. https://doi.org/10.1029/TR039i004p00669
39. McCuen RH (1998) Hydrologic analysis and design, 4th edn. Pearson
40. Mondal GC, Singh AK, Singh TB (2018) Damodar river basin: storehouse of Indian coal, pp 259–272
41. Moore ID, Burch GJ (1986) Physical basis of the length-slope factor in the universal soil loss equation. Soil Sci Soc Am J 50:1294–1298. https://doi.org/10.2136/sssaj1986.03615995005000050042x
42. Morgan RPC (2009) Soil erosion & conservation
43. Nearing MA, Foster GR, Lane LJ, Finkner SC (1989) A process-based soil erosion model for USDA—water erosion prediction project technology. Trans ASAE 32:1587–1593
44. O'Farrell PJ, De Lange WJ, Le Maitre DC et al (2011) The possibilities and pitfalls presented by a pragmatic approach to ecosystem service valuation in an arid biodiversity hotspot. J Arid Environ 75:612–623. https://doi.org/10.1016/j.jaridenv.2011.01.005
45. Pimentel D, Harvey C, Resosudarmo P et al (1995) Environmental and economic costs of soil erosion and conservation benefits. Science (80-) 267:1117–1123. https://doi.org/10.1126/science.267.5201.1117
46. Rajbanshi J, Bhattacharya S (2022) Modelling the impact of climate change on soil erosion and sediment yield: a case study in a sub-tropical catchment, India. Model Earth Syst Environ 8:689–711. https://doi.org/10.1007/s40808-021-01117-4
47. Rajbanshi J, Bhattacharya S (2020) Assessment of soil erosion, sediment yield and basin specific controlling factors using RUSLE-SDR and PLSR approach in Konar river basin. India. J Hydrol 587:124935. https://doi.org/10.1016/j.jhydrol.2020.124935
48. Renard KG, Foster GR, Weesies GA, Porter JP (1991) RUSLE: revised universal soil loss equation. J Soil Water Conserv 46:30–33
49. Renard KG et al (1997) Predicting soil erosion by water: a guide to conservation planning with the revised universal soil loss equation (RUSLE). US Department of Agriculture, Agriculture Handbook No.703USDA, USDA, Washington DC
50. Sengupta S, Mohinuddin S, Arif M (2022) Sub-watershed prioritization for soil erosion potentiality estimation in Tenughat catchment, India. Geocarto Int 37:9160–9189. https://doi.org/10.1080/10106049.2021.2017008
51. Singh G, Babu R, Narain P et al (1981) Soil loss prediction research in India. Cent Soil Water Conserv Res Train Institute, Dehradun Steel
52. Tarboton DG (2014) TauDem 5.3.7: terrain analysis using digital elevation models. http://hydrology.usu.edu/taudem/taudem5/downloads.html
53. Thomas J, Joseph S, Thrivikramji KP (2018) Estimation of soil erosion in a rain shadow river basin in the southern Western Ghats, India using RUSLE and transport limited sediment delivery function. Int Soil Water Conserv Res 6:111–122. https://doi.org/10.1016/j.iswcr.2017.12.001
54. USDA Handbook no. 282 (1981) USDA agriculture handbook
55. van der Knijff JM, Jones RJA, Montanarella L (2000) Soil erosion risk assessment in Europe. Luxemb Off Off Publ Eur Communities EUR 19022:32
56. Walling DE (1983) The sediment delivery problem. J Hydrol 65:209–237. https://doi.org/10.1016/0022-1694(83)90217-2
57. Wang L, Li Y, Gan Y et al (2024) Rainfall erosivity index for monitoring global soil erosion. CATENA 234:107593. https://doi.org/10.1016/j.catena.2023.107593
58. Wessels KJ, Prince SD, Malherbe J et al (2007) Can human-induced land degradation be distinguished from the effects of rainfall variability? A case study in South Africa. J Arid Environ 68:271–297. https://doi.org/10.1016/j.jaridenv.2006.05.015
59. Williams JR (1977) Sediment delivery ratios determined with sediment and runoff models. IAHS Publ 168–179
60. Wischmeier WH, Smith DD (1978) USLE Predicting rainfall erosion lossses: a guide to conservation planning

61. Wischmeier WH, Smith DD (1965) Rainfall-erosion losses from cropland in cooperation with purdue agricultural experiment station. Agric Handb no 282 US Department, vol 49
62. Wuepper D, Borrelli P, Finger R (2019) Countries and the global rate of soil erosion. Nat Sustain 3:51–55. https://doi.org/10.1038/s41893-019-0438-4
63. Yan R, Zhang X, Yan S, Chen H (2018) Estimating soil erosion response to land use/cover change in a catchment of the Loess Plateau, China. Int Soil Water Conserv Res 6:13–22. https://doi.org/10.1016/j.iswcr.2017.12.002
64. Yang D, Kanae S, Oki T et al (2003) Global potential soil erosion with reference to land use and climate changes. Hydrol Process 17:2913–2928. https://doi.org/10.1002/hyp.1441
65. Zerihun M, Mohammedyasin MS, Sewnet D et al (2018) Assessment of soil erosion using RUSLE, GIS and remote sensing in NW Ethiopia. Geoderma Reg 12:83–90. https://doi.org/10.1016/j.geodrs.2018.01.002

Chapter 5
Sediment Connectivity Index

Abstract This chapter focuses on the Sediment Connectivity Index (SCI) to assess sediment transport dynamics in the Damodar River Basin (DRB), with an emphasis on sub-basin-wise connectivity. The research calculates the SCI for each sub-basin to quantify the efficiency of sediment flow from erosion hotspots to sediment deposition zones within the basin. The study highlights important places that affect sediment transportation and dynamics by analyzing the interaction between Soil Erosion Hotspots (SEH) and the Connectivity Index (IC). The findings reveal the most connective sub-basins within the DRB, indicating areas with high sediment delivery potential and strong linkage to downstream sedimentation. Understanding these connections is essential for targeting soil conservation efforts and improving sediment management strategies in the basin. This chapter provides valuable insights for optimizing watershed management, mitigating erosion, and enhancing the overall health and sustainability of the Damodar River Basin's hydrological systems.

Keywords Sediment connectivity index · Dis(connectivity) · Landscape · Sub-basin · Damodar River basin · Sediment delivery potential · Linkage · Sediment pathways

5.1 Introduction

The concept of 'Connectivity' is increasingly being employed in hydrology, sedimentology, and landscape science [12]. The connectivity of the catchment or landscape system refers to the ease of flow of energy and matter throughout the ecosystem or fluvial systems. Landscape connectivity is a primary control of catchment-scale fluxes of water and sediment [12]. Connectivity is an emergent substance of a geomorphic system or fluvial system [30]. The catchment exhibits several types of connectivity: (1) Landscape connectivity, (2) Hydrological connectivity, and (3) Sedimentological connectivity. The majority of the Damodar River catchment is located in the upper and middle regions, with all materials flowing through the funnel-shaped lower catchment. The construction of various large and small dams in the upper

S. A. Alam and R. Maiti, *Sediment Source to Sink: Deciphering Sediment Connectivity to Large Dams in Damodar River Basin*, Earth and Environmental Sciences Library,
https://doi.org/10.1007/978-3-031-87072-9_5

and middle catchments of the DRB has disrupted hydrological and sedimentological connectivity. These dams have faced significant sedimentation and siltation issues, reducing their water-holding capacity and accelerating their decline into ineffective structures. Therefore, the sediment connectivity, or Sediment Connectivity Index (SCI), becomes crucial in understanding the dynamics of sediment transport from the upper catchment to the main channel and into dams or reservoirs. Sediment connectivity is positively correlated with sediment yield in the fluvial system. Numerous scholars have conducted research on hydrological connectivity, sediment connectivity, and landscape connectivity across the globe [1, 3–6, 8, 12, 14–16, 19, 21, 23, 25–27, 29–31, 33].

In this study, the Sediment Connectivity Index (SCI) or Index of Connectivity (IC) is applied to analyze spatial sediment dynamics in the Damodar catchment. The study focuses on three major dams—Maithon, Panchet, and Tenughat—to understand sedimentation and siltation issues originating from the upper catchment of the Damodar River Basin (DRB). The contrasting geomorphology of the DRB influences sediment transfer processes with varying intensity, leading to differences in SCI across these dam catchments. Anthropogenic activities exacerbate the rate at which sediment is transported from the upper catchments to the reservoir. This sediment travels through connectivity pathways, and these act as one such conduit between the sediment sources and the reservoir.

5.2 Conceptual Background of Connectivity Index (IC)

The connectivity of the catchment or landscape system refers to the efficiency in the transfer of energy, and matter throughout the ecosystem or fluvial systems. Many geomorphological studies have found that the primary process of sediment transfer from the source to the channel network determines the landscape evaluation [18, 19]. In a particular fluvial system, the high mass transfer efficiency is also characterized by a significant degree of connectivity [2, 18, 19]. Various algorithms have been applied to estimate sediment delivery issues and sediment yield. Both the upslope and downslope approaches have been considered in this analysis to better understand the sediment transport dynamics [5, 6, 10, 19, 20, 24] for developing the index of connectivity (IC) approach. The GIS-based approach for developing the Index of Connectivity (IC) is demonstrated through a straightforward illustration (Fig. 5.1). This method helps to visually represent the sediment connectivity across the catchment.

A conceptual framework for sediment connectivity was developed by [5], which incorporates various types of landscape linkages across different spatial scales within the catchment area. Landscape connectivity is a primary control of catchment-scale fluxes of water and sediment [12]. The conceptual model for the sediment (dis)connectivity framework was meticulously developed by integrating longitudinal, lateral, and vertical blockages [13], providing a comprehensive approach to

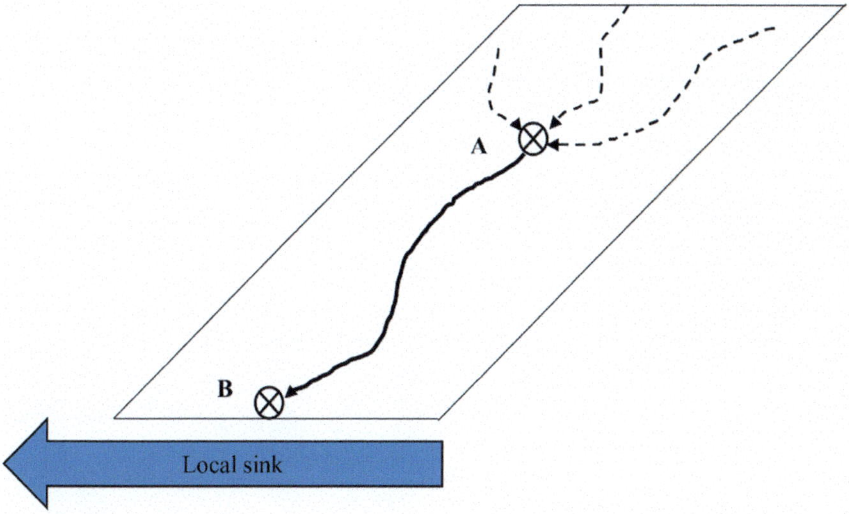

Fig. 5.1 Simple slope with flow path between the source area A and the local sink B [1]

understanding multi-dimensional sediment flow impediments. The Index of Connectivity (IC), also known as the Sediment Connectivity Index (SCI), was applied to the Damodar catchment based on the methodology proposed by Borselli et al. [1] and subsequently modified by [6]. The conceptual framework for estimating the Index of Connectivity (IC) was illustrated by [6], as depicted in Fig. 5.2.

5.2.1 Methods

The Sediment Connectivity Index (SCI) represents the probability of sediment transfer from a given location within a catchment, reflecting the integrated connectivity potential across the landscape [29]. Sediment connectivity assesses each sediment source to evaluate the extent of hillslope-to-channel transfer of water and sediment. This study employed a GIS-based approach developed by [6], building on the foundational work by Borselli et al. [1]. The Connectivity Index (IC) was derived using the following methodology as proposed by Borselli et al. [1]:

$$IC_i = \log_{10}\left(\frac{D_{up}}{D_{dn}}\right) \tag{5.1}$$

Also, Eq. (5.1) can be written as follows:

$$IC_i = \log_{10}\left(\frac{\overline{W}\,\overline{S}\sqrt{A}}{\sum_i \frac{d_i}{W_i S_i}}\right) \tag{5.2}$$

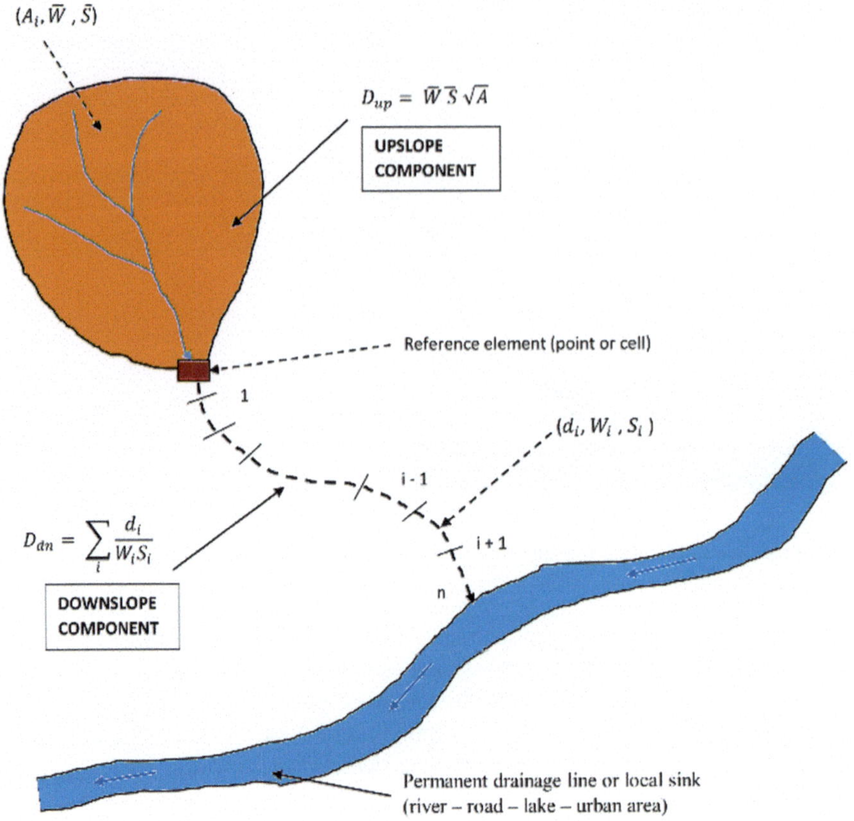

Fig. 5.2 Definition of upslope and downslope component of the index of connectivity [1, 6]

where,

IC_i Index of sediment connectivity
D_{up} Upslope component of connectivity
D_{dn} Downslope component of connectivity.

The upslope component of connectivity (D_{up}) is the downward routing potentiality of sediment that emanated from upslope and is estimated as follows:

$$D_{up} = \overline{WS}\sqrt{A} \tag{5.3}$$

where,

$\overline{W}$ The average weighting factor of the upslope contributing area (dimensionless)
$\overline{S}$ The average slope gradient of the upslope contributing area (m/m)
$\sqrt{A}$ Upslope contributing area (m^2).

The average weighting factor of the upslope area ($\overline{W}$) estimated as follows:

$$W = 1 - \left(\frac{RI}{RI_{MAX}}\right) \quad (5.3a)$$

where the RI indicates the Roughness Index and this is estimated as follows:

$$RI = \sqrt{\frac{\sum_{i=1}^{25}(X_i - X_m)^2}{25}} \quad (5.3b)$$

where,

RI	Roughness index
RI_{MAX}	Maximum value of RI in the study area
X_i	value of one specific cell of the residual topography within the moving window
X_m	Mean of 25 cell value
25	Number of processing cells among the 5 × 5 moving window.

The downslope component of connectivity (D_{dn}) indicates the flow path length of that individual particle that has to transit to arrive at the nearest target or sink. Therefore, D_{dn} This can be expressed as:

$$D_{dn} = \sum_i \frac{d_i}{W_i S_i} \quad (5.4)$$

where, d_i is the length of the flow path along the *i*th cell according to the steepest downslope direction (m), W_i and S_i are the weighting factor and the slope gradient of the *i*th cell, respectively. Therefore, in this study, the different years' IC maps were prepared by Eq. (5.1) in the Damodar River basin (DRB) for the management of the sedimentation and siltation problems in major dams.

5.3 Sediment Connectivity Index (IC) of Damodar River Basin (DRB)

The spatial sediment connectivity was carried out to understand the matter and energy transfer pathways from the source to the sink for the years 1990, 2005, and 2021. The basin experiences significant human encroachment, including mining, agriculture, and deforestation, which accelerates sediment production from the upper catchment. However, sediment connectivity throughout the Damodar River basin is heavily regulated by various major dams, influencing sediment flow from the upper to lower catchment areas. Thus, understanding the dynamics of sediment connectivity is essential for addressing sediment yield and sedimentation challenges across all major dams within this catchment for various years.

5.3.1 Year 1990

The Index of Connectivity (IC) map for 1990 reveals that IC values range from a minimum of − 5.54 to a maximum of 3.81 (Fig. 5.3). In this map, higher IC values are concentrated near the main channel, while lower values are distributed along hillslopes. The lower catchment of the Damodar basin exhibits minimal connectivity, indicating weaker sediment connections with the main channel, which suggests sediment deposition in adjacent areas. In contrast, the upper and middle catchments display strong sediment connectivity to the main channels, with the highest connectivity observed in the main channel of the Damodar River. The mean sediment connectivity index (SCI) for 1990 was − 1.27 (SD = 0.84). Sediment connectivity in the Damodar catchment was categorized into five classes: Very High, High, Moderate, Low, and Very Low. Approximately 34.12% of the area showed high to very high SCI, while 33.32% displayed low to very low SCI for 1990 (Fig. 5.3).

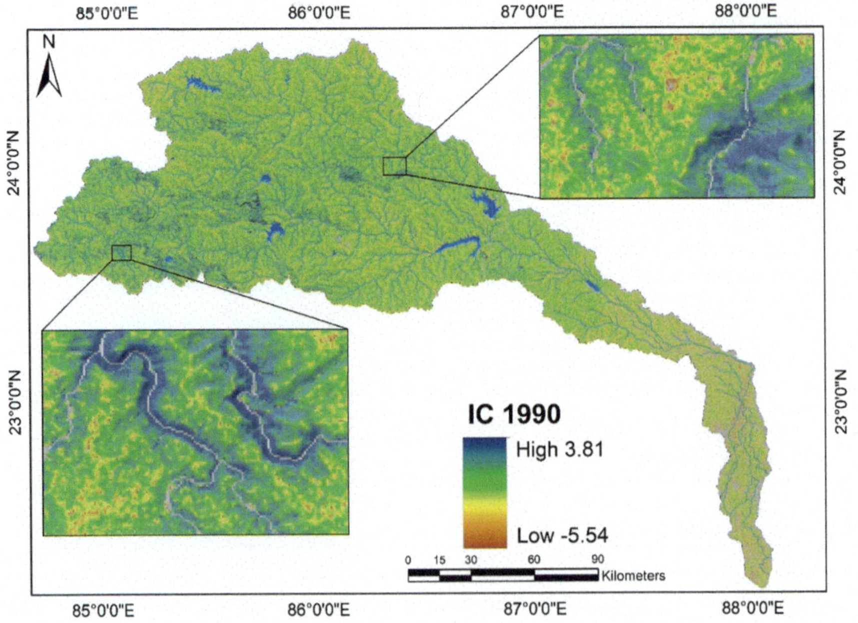

Fig. 5.3 Index of connectivity (IC) or sediment connectivity index (SCI) map of Damodar River basin (DRB) for the year 1990

5.3.2 *Year 2005*

The Connectivity Index (IC) map for 2005 reveals spatial variations in sediment connectivity, with values ranging from a minimum of − 6.33 to a maximum of 3.59 (Fig. 5.4). Higher IC values are concentrated along the main channel, whereas lower values usually appear in places farther away from the channels. In the lower Damodar basin, minimal connectivity values suggest relatively weak sediment linkage with the main channel, implying sediment deposition in adjacent areas. Conversely, the upper and middle catchments exhibit more consistent sediment connectivity in the hillslope areas to main channels, with peak connectivity observed along the primary course of the Damodar River. The mean sediment connectivity index (SCI) for 2005 was − 1.26 (SD = 0.86). SCI values in the Damodar catchment were categorized into five classes: Very High, High, Moderate, Low, and Very Low, with over 37.22% of the area showing high to very high connectivity and 31.08% showing low to very low connectivity for 2005 (Fig. 5.4).

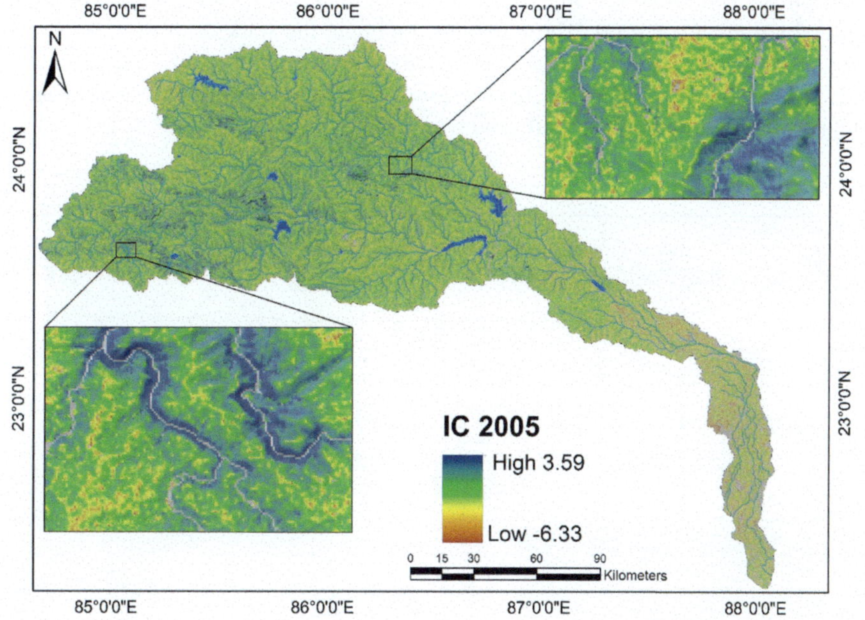

Fig. 5.4 Index of connectivity (IC) or sediment connectivity index (SCI) map of Damodar River basin (DRB) for the year 2005

5.3.3 Year 2021

The **Index of Connectivity (IC)** analysis for the Damodar River basin in 2021 reveals significant spatial variability in sediment connectivity, with IC values ranging from − 5.15 to 3.80 (Fig. 5.5). Higher IC values are concentrated near the main channels, indicating stronger sediment connectivity and efficient sediment transfer pathways. Conversely, lower IC values are observed in areas farther from the main channels, particularly in the lower catchment, where weaker sediment transport potential suggests deposition in surrounding areas due to flat topography and reduced hydrological activity. The lower catchment's reduced connectivity highlights sediment deposition zones, while the upper and middle catchments exhibit consistently higher IC values. The mean Sediment Connectivity Index (SCI) for 2021 was − 1.28 (SD = 0.86). Sediment connectivity is categorized into five classes: Very High, High, Moderate, Low, and Very Low. Approximately 34.36% of the basin area falls within the High to Very High connectivity classes, predominantly located near channels and in the foothills of the hilly regions within the upper catchment of the DRB. In contrast, 33.13% of the area exhibits Low to Very Low connectivity, indicating limited sediment transport. These findings highlight the stark contrast in sediment dynamics across the basin. High-connectivity zones are highly susceptible to soil erosion and actively contribute to downstream sedimentation, which significantly impacts reservoir storage capacity and river systems.

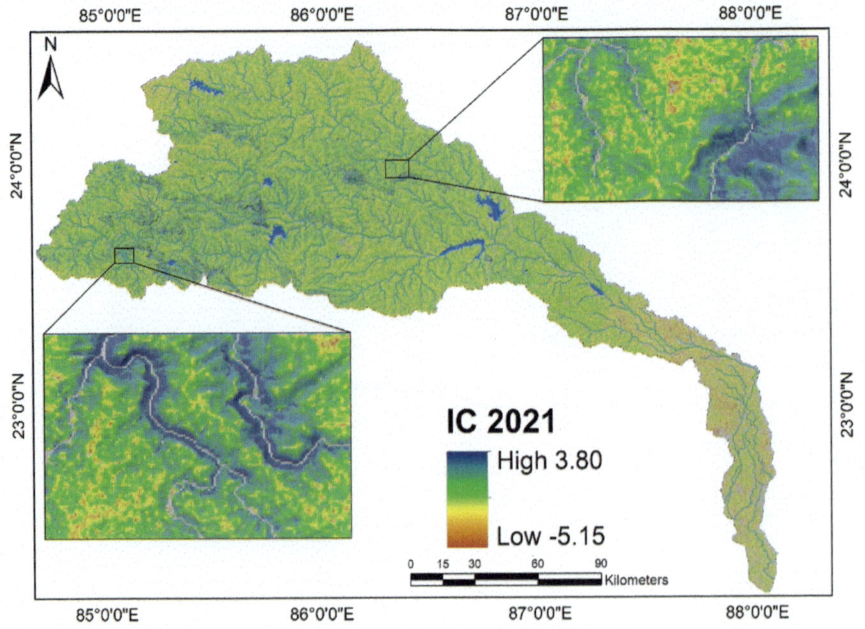

Fig. 5.5 Index of connectivity (IC) or sediment connectivity index (SCI) map of Damodar River basin (DRB) for the year 2021

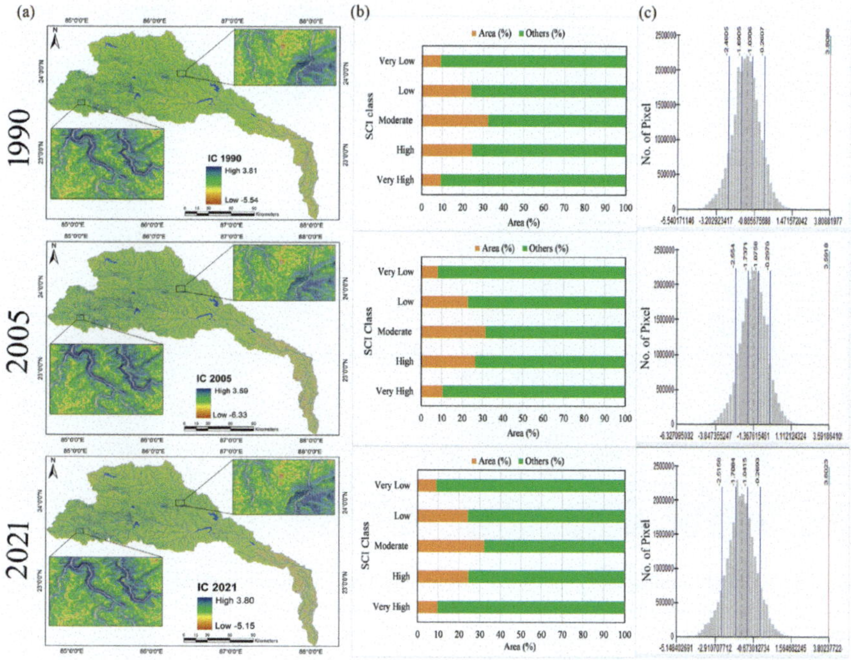

Fig. 5.6 The map shows the **a** Sediment connectivity index (SCI) map, **b** Percentage of areal distribution in different classes (very high to very low) of SCI, and **c** The histogram shows the distribution of SCI values for the year of 1990, 2005 and 2021

The catchment IC values are identical throughout the years of 1990, 2005, and 2021. The histograms for the 1990, 2005, and 2021 maps display a positively skewed, non-normal distribution for the entire Damodar catchment. High sediment connectivity classes accounted for 9.46% of the Damodar River Basin (DRB) in 1990, 10.65% in 2005, and 9.78% in 2021 (Fig. 5.6). The plateau areas in the Damodar catchment consistently exhibit SCI values above the mean across all years, while the lower catchment shows SCI values below the mean, though only the main channel network demonstrates strong connectivity throughout these years (Fig. 5.7).

5.4 Sub-Basin-Wise Sediment Connectivity Index of DRB

Connectivity Index (SCI) values were calculated for 1990, 2005, and 2021 against all 23 sub-basins. The average sediment connectivity values for these years are depicted in Fig. 5.8. Sub-basins SB6, SB7, and SB12 exhibit very high sediment connectivity in both 1990 and 2021, while SB6, SB7, SB12, and SB19 demonstrate very high connectivity for the year 2005. In contrast, sub-basins SB22 and SB23 are characterized by very low sediment connectivity. Sub-basins SB2, SB8, SB9, SB10, SB11,

Fig. 5.7 Sediment connectivity pathways (well-connected) with steep-sided residual hills with extensively weathered materials produced massive sediment towards patratu lake (sink) (Photo credit: Farhina Akhtar)

SB15, and SB19 display high sediment connectivity for both 1990 and 2021, whereas SB2, SB4, SB5, SB8, SB9, SB10, SB11, and SB15 exhibit high connectivity specifically for 2005. The correlation between the SCI and sediment yield (SY) across all sub-basins in the Damodar catchment shows R^2 values of 0.73 for 1990, 0.60 for 2005, and 0.94 for 2021. The sub-basin SB7 has the highest sediment output, displaying maximum connectivity values consistently across all years investigated (1990, 2005, and 2021).

5.5 Linkage Between Soil Erosion Hotspot (SEH) and Connectivity Index (IC)

In the upper and middle catchments of the DRB, soil erosion hotspots (SEH) and the index of connectivity (IC) have been analyzed for the years 1990, 2005, and 2021. Over the specified period, the number of SEHs in the Maithon dam catchment fluctuated, recording 18 in 1990, 22 in 2005, and 16 in 2021. For the Panchet dam, the SEH counts were 27, 16, and 16 for the respective years, while the Tenughat dam exhibited 25, 16, and 13 SEH during the same timeframe. Notably, the total area of SEHs increased across all three dams from 1990 to 2021, expanding from 449.60 to 453.97 km^2 in the Maithon dam, from 669.70 to 705.93 km^2 in the Panchet dam, and from 1077.19 to 1264.94 km^2 in the Tenughat dam. Despite the increase in SEH areas, the overall number of hotspots decreased between 1990 and 2021 for

- Sub-basins SB6, SB7, and SB12 demonstrated very high sediment connectivity (SCI) in both 1990 and 2021, while SB19 exhibited very high connectivity levels in 2005 in addition to SB6, SB7, and SB12.
- Sub-basins SB22 and SB23 consistently displayed very low sediment connectivity (SCI) across all the years (1990, 2005, and 2021).
- A positive correlation is observed between SCI and SY across all sub-basins in the Damodar River Basin for all years (1990 = 0.73, 2005 = 0.60, and 2021 = 0.94), indicating an increased likelihood of sediment being transported to the channel.
- The number of soil erosion hotspots (SEHs) fluctuated from 1990 to 2021. The numbers of SEH for the Maithon dam were 18, 22, and 16 in 1990, 2005, and 2021, and the same became 27, 16, and 16 for the Panchet Dam and 25, 16, and 13 for the Tenughat dam in 1990, 2005, and 2021 respectively.
- The area of SEH increased from 449.60 to 453.97 km^2 for the Maithon dam, from 669.70 to 705.93 km^2 for the Panchet Dam, and from 1077.19 to 1264.94 km^2 for the Tenughat Dam between 1990 and 2021.
- Although the number of hotspots has declined since 1990 across all dams (Maithon, Panchet, and Tenughat), the total area of these hotspots has increased, indicating a high rate of spatial expansion and coalescence of hotspots.
- Notably, very good sediment connectivity (SCI) was consistently observed in the sub-basins SB6, SB7, SB8, SB11, SB12, and SB19 across the years.

5.8 Conclusion

This chapter examines the temporal variations in sediment connectivity (SCI) within the Damodar River Basin (DRB) for the years 1990, 2005, and 2021. The connectivity is coupled with the RUSLE model factor to evaluate the sedimentation effect of major dams (Maithon, Panchet, and Tenughat) in the Damodar catchment. The sediment connectivity within the catchment areas of major dams is correlated with the sub-basins to identify the most significant sub-basins that supply the maximum amount of sediment to the mainstream channel as well as major dams (Maithon, Panchet, and Tenughat). The increasing trend of sediment connectivity (SCI) is found from 1990 to 2021. The significant sub-basins were identified based on coupling with the soil erosion hotspot (SEH), sediment connectivity (SCI), and soil erosion probability (SE). The sub-basins SB6, SB7, and SB12 are identified as significant, highly connected soil erosion hotspots (SEH) to the mainstream network and major dams. The sediment connectivity is high in the upper catchment of DRB, Whereas the lower catchment of the Damodar Basin is characterized by plain land and very poor connectivity.

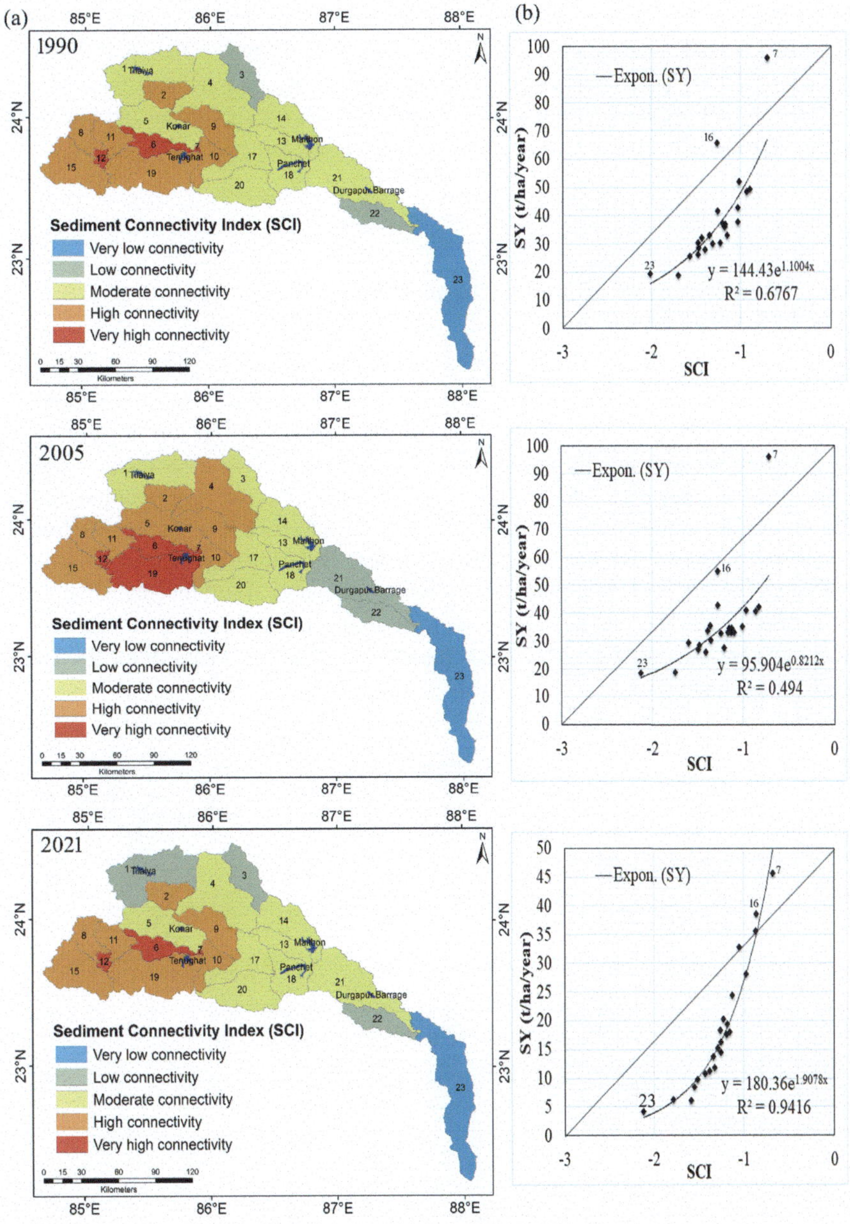

Fig. 5.8 The map showing the (a) Mean sediment connectivity index (SCI) map of all sub-basins in Damodar catchment and (b) the Relationship between the SCI and SY of all sub-basins in Damodar catchment for the year 1990, 2005, and 2021 consequently

all three dams. This pattern signifies that the areas at risk of erosion are expanding, whereas the particular locations of hotspots are diminishing as the spatial spread leads neighboring hotspots to merge. The presence of soil erosion hotspots combined with strong connectivity throughout the dams indicates a significant likelihood of spreading and efficient sediment input to the mainstream network and the dams themselves.

5.6 Identification of Most Connective Sub-Basins

Identifying sediment source zones [7, 9, 11, 28] in the upper catchment of a dam is essential for maintaining reservoir storage capacity, conserving water quality, and implementing efficient erosion control measures. Soil erosion hotspots [17, 22, 32] that are statistically significant ($p < 0.05$) and have a sediment connectivity value greater than 1.5 (IC > 1.5) are referred to as sediment source zones. The Damodar catchment has been divided into 23 sub-basins, where the Sediment Connectivity Index (SCI) has been utilized to identify the most connected sub-basins that contribute the highest probability of sediment input to the main channel and the dams (Fig. 5.9). The overall mean sediment connectivity for the entire Damodar catchment was recorded as − 1.27, − 1.26, and − 1.28 for the years 1990, 2005, and 2021, respectively.

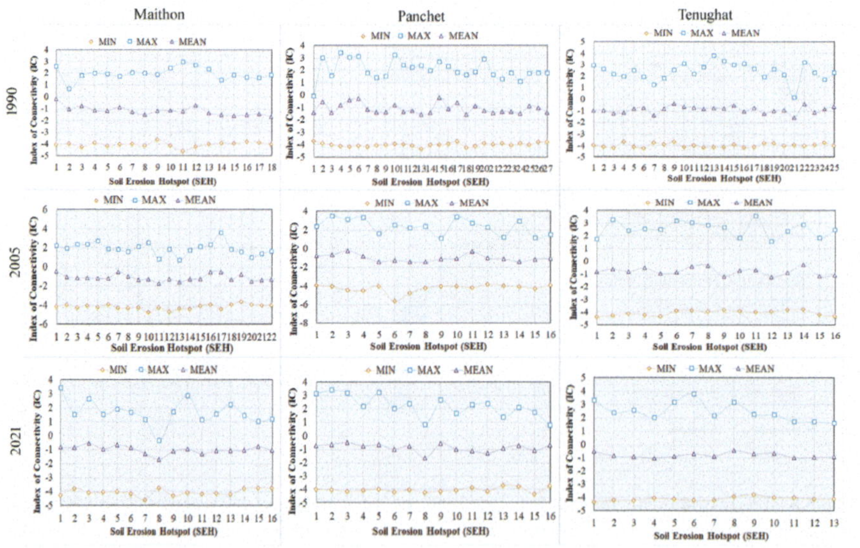

Fig. 5.9 The number of soil erosion hotspots (SEH) and the variation of Index of connectivity (IC) values within the various major dams in Damodar catchment (a) Maithon, (b) Panchet, and (c) Tenughat for the years 1990, 2005 and 2021

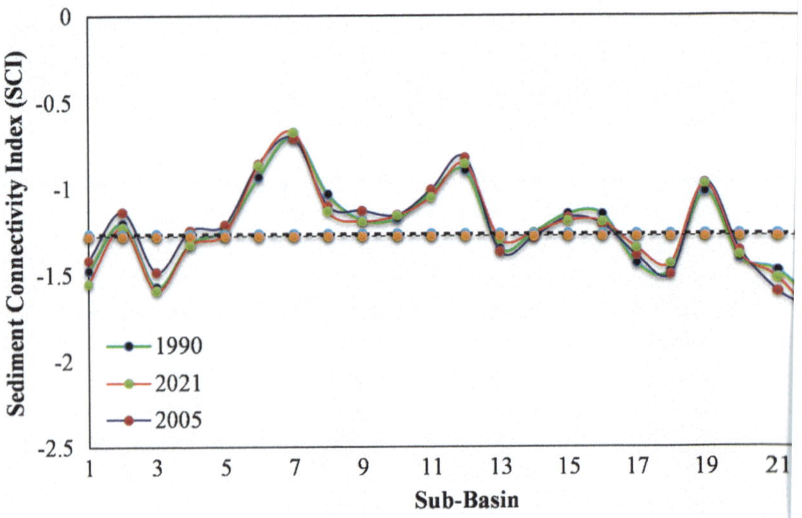

Fig. 5.10 Sub-basins-wise mean sediment connectivity index (SCI) compared to mean S

Sub-basin-wise estimates of mean sediment connectivity were calculated years 1990, 2005, and 2021, revealing that sub-basins SB6, SB7, SB8, SB11 and SB19 exhibited very good sediment connectivity across all years, demon values above the overall average for the Damodar catchment. Conversely, sub SB1, SB3, SB18, SB21, SB22, and SB23 consistently showed very low se connectivity, falling below the average for the catchment (Fig. 5.10). Notab basin SB7 recorded significant sediment connectivity with values of 0.7 in 19 in 2005, and 0.82 in 2021. Overall, sub-basins exhibiting a high sediment conn class showed an increasing trend from 1990 to 2021, while those with low se connectivity demonstrated a decreasing trend over the same period.

5.7 Major Findings

The findings in this chapter are as follows.

- The mean Sediment Connectivity Index (SCI) values were − 1.27 (SD = − 1.26 (SD = 0.86), and − 1.28 (SD = 0.86) for the years 1990, 2005, an respectively.
- In 1990, over 34.12% of the area exhibited very high to high sediment conn (SCI), while this figure rose to 37.22% in 2005 and slightly decreased to in 2021.
- Conversely, more than 33.32% of the area exhibited very low to low se connectivity (SCI) in 1990, decreased to 31.08% in 2005, and then in again to 33.13% in 2021.

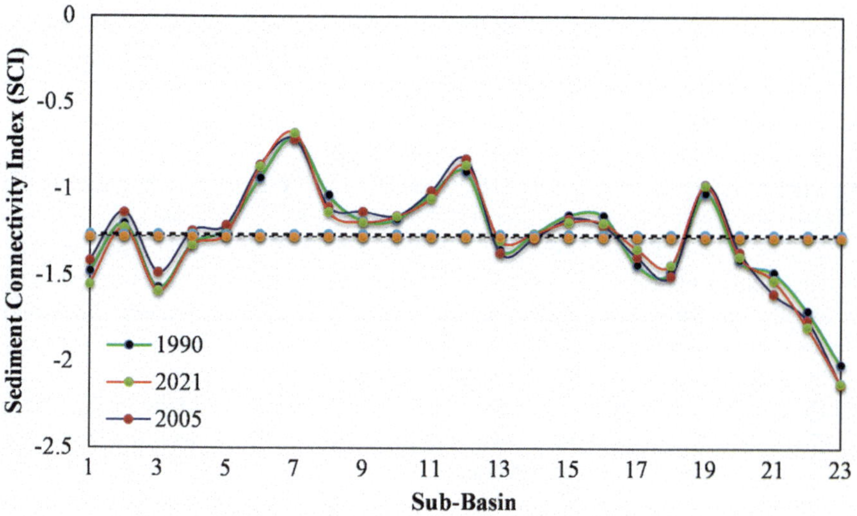

Fig. 5.10 Sub-basins-wise mean sediment connectivity index (SCI) compared to mean SCI

Sub-basin-wise estimates of mean sediment connectivity were calculated for the years 1990, 2005, and 2021, revealing that sub-basins SB6, SB7, SB8, SB11, SB12, and SB19 exhibited very good sediment connectivity across all years, demonstrating values above the overall average for the Damodar catchment. Conversely, sub-basins SB1, SB3, SB18, SB21, SB22, and SB23 consistently showed very low sediment connectivity, falling below the average for the catchment (Fig. 5.10). Notably, sub-basin SB7 recorded significant sediment connectivity with values of 0.7 in 1990, 0.72 in 2005, and 0.82 in 2021. Overall, sub-basins exhibiting a high sediment connectivity class showed an increasing trend from 1990 to 2021, while those with low sediment connectivity demonstrated a decreasing trend over the same period.

5.7 Major Findings

The findings in this chapter are as follows.

- The mean Sediment Connectivity Index (SCI) values were − 1.27 (SD = 0.84), − 1.26 (SD = 0.86), and − 1.28 (SD = 0.86) for the years 1990, 2005, and 2021, respectively.
- In 1990, over 34.12% of the area exhibited very high to high sediment connectivity (SCI), while this figure rose to 37.22% in 2005 and slightly decreased to 34.36% in 2021.
- Conversely, more than 33.32% of the area exhibited very low to low sediment connectivity (SCI) in 1990, decreased to 31.08% in 2005, and then increased again to 33.13% in 2021.

- Sub-basins SB6, SB7, and SB12 demonstrated very high sediment connectivity (SCI) in both 1990 and 2021, while SB19 exhibited very high connectivity levels in 2005 in addition to SB6, SB7, and SB12.
- Sub-basins SB22 and SB23 consistently displayed very low sediment connectivity (SCI) across all the years (1990, 2005, and 2021).
- A positive correlation is observed between SCI and SY across all sub-basins in the Damodar River Basin for all years (1990 = 0.73, 2005 = 0.60, and 2021 = 0.94), indicating an increased likelihood of sediment being transported to the channel.
- The number of soil erosion hotspots (SEHs) fluctuated from 1990 to 2021. The numbers of SEH for the Maithon dam were 18, 22, and 16 in 1990, 2005, and 2021, and the same became 27, 16, and 16 for the Panchet Dam and 25, 16, and 13 for the Tenughat dam in 1990, 2005, and 2021 respectively.
- The area of SEH increased from 449.60 to 453.97 km^2 for the Maithon dam, from 669.70 to 705.93 km^2 for the Panchet Dam, and from 1077.19 to 1264.94 km^2 for the Tenughat Dam between 1990 and 2021.
- Although the number of hotspots has declined since 1990 across all dams (Maithon, Panchet, and Tenughat), the total area of these hotspots has increased, indicating a high rate of spatial expansion and coalescence of hotspots.
- Notably, very good sediment connectivity (SCI) was consistently observed in the sub-basins SB6, SB7, SB8, SB11, SB12, and SB19 across the years.

5.8 Conclusion

This chapter examines the temporal variations in sediment connectivity (SCI) within the Damodar River Basin (DRB) for the years 1990, 2005, and 2021. The connectivity is coupled with the RUSLE model factor to evaluate the sedimentation effect of major dams (Maithon, Panchet, and Tenughat) in the Damodar catchment. The sediment connectivity within the catchment areas of major dams is correlated with the sub-basins to identify the most significant sub-basins that supply the maximum amount of sediment to the mainstream channel as well as major dams (Maithon, Panchet, and Tenughat). The increasing trend of sediment connectivity (SCI) is found from 1990 to 2021. The significant sub-basins were identified based on coupling with the soil erosion hotspot (SEH), sediment connectivity (SCI), and soil erosion probability (SE). The sub-basins SB6, SB7, and SB12 are identified as significant, highly connected soil erosion hotspots (SEH) to the mainstream network and major dams. The sediment connectivity is high in the upper catchment of DRB, Whereas the lower catchment of the Damodar Basin is characterized by plain land and very poor connectivity.

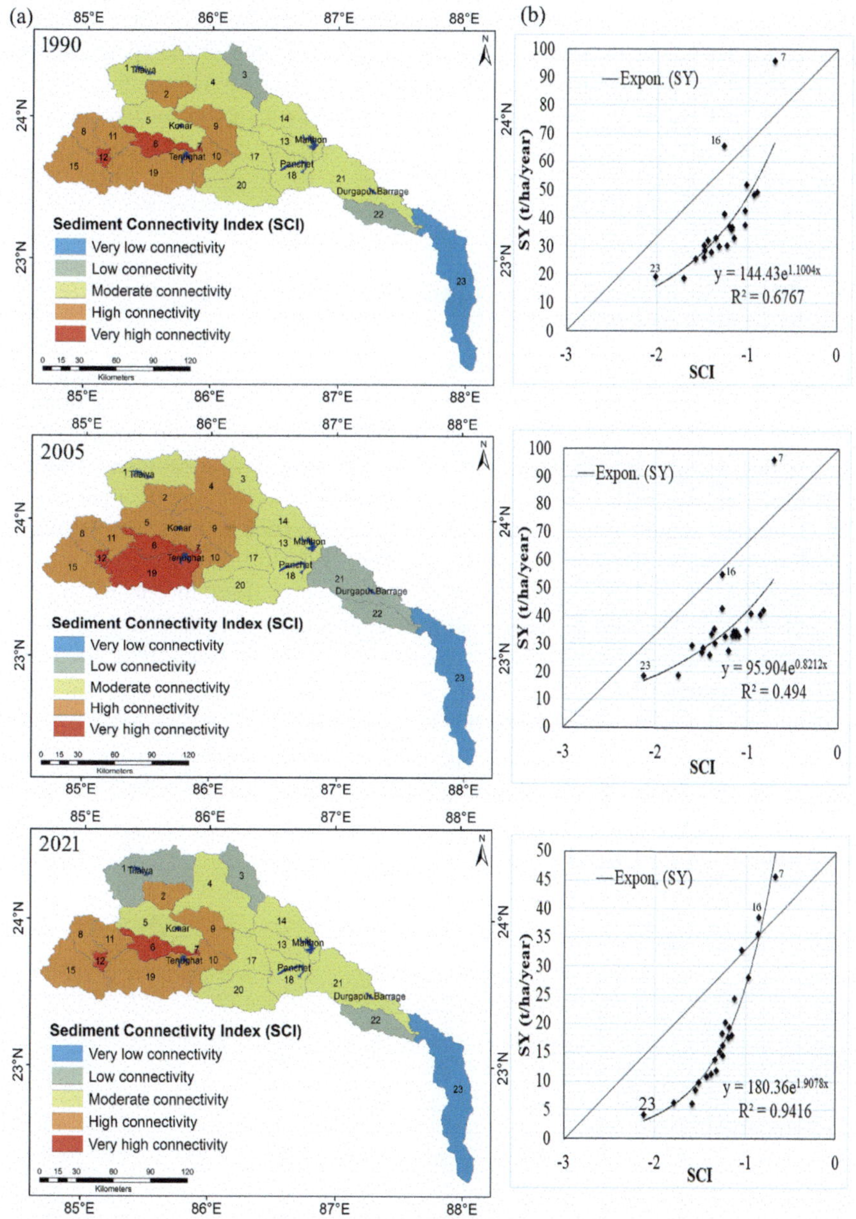

Fig. 5.8 The map showing the (a) Mean sediment connectivity index (SCI) map of all sub-basins in Damodar catchment and (b) the Relationship between the SCI and SY of all sub-basins in Damodar catchment for the year 1990, 2005, and 2021 consequently

all three dams. This pattern signifies that the areas at risk of erosion are expanding, whereas the particular locations of hotspots are diminishing as the spatial spread leads neighboring hotspots to merge. The presence of soil erosion hotspots combined with strong connectivity throughout the dams indicates a significant likelihood of spreading and efficient sediment input to the mainstream network and the dams themselves.

5.6 Identification of Most Connective Sub-Basins

Identifying sediment source zones [7, 9, 11, 28] in the upper catchment of a dam is essential for maintaining reservoir storage capacity, conserving water quality, and implementing efficient erosion control measures. Soil erosion hotspots [17, 22, 32] that are statistically significant ($p < 0.05$) and have a sediment connectivity value greater than 1.5 (IC > 1.5) are referred to as sediment source zones. The Damodar catchment has been divided into 23 sub-basins, where the Sediment Connectivity Index (SCI) has been utilized to identify the most connected sub-basins that contribute the highest probability of sediment input to the main channel and the dams (Fig. 5.9). The overall mean sediment connectivity for the entire Damodar catchment was recorded as − 1.27, − 1.26, and − 1.28 for the years 1990, 2005, and 2021, respectively.

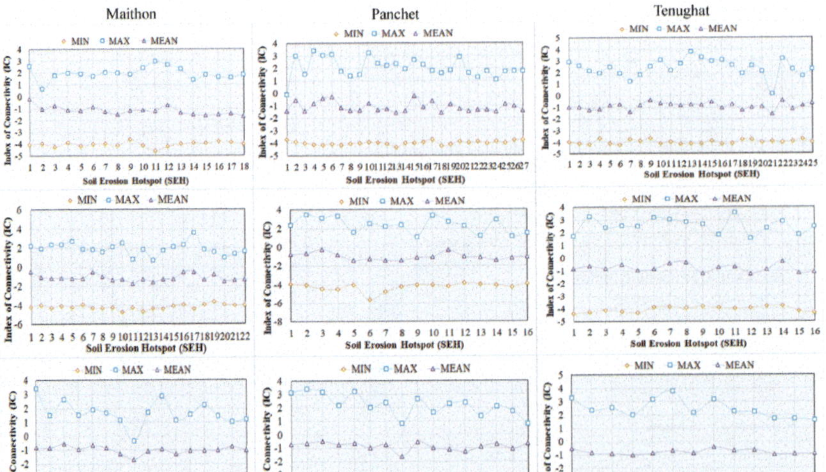

Fig. 5.9 The number of soil erosion hotspots (SEH) and the variation of Index of connectivity (IC) values within the various major dams in Damodar catchment (a) Maithon, (b) Panchet, and (c) Tenughat for the years 1990, 2005 and 2021

References

1. Borselli L, Cassi P, Torri D (2008) Prolegomena to sediment and flow connectivity in the landscape: a GIS and field numerical assessment. CATENA 75(3):268–277. https://doi.org/10.1016/j.catena.2008.07.006
2. Bracken LJ, Croke J (2007) The concept of hydrological connectivity and its contribution to understanding runoff-dominated geomorphic systems. Hydrol Process 21(13):1749–1763. https://doi.org/10.1002/hyp.6313
3. Bracken LJ, Turnbull L, Wainwright J, Bogaart P (2015) Sediment connectivity: a framework for understanding sediment transfer at multiple scales. Earth Surf Proc Land 40(2):177–188. https://doi.org/10.1002/esp.3635
4. Bracken LJ, Wainwright J, Ali GA, Tetzlaff D, Smith MW, Reaney SM, Roy AG (2013) Concepts of hydrological connectivity: research approaches, pathways and future agendas. Earth Sci Rev 119:17–34. https://doi.org/10.1016/j.earscirev.2013.02.001
5. Brierley G, Fryirs K, Jain V (2006) Landscape connectivity: the geographic basis of geomorphic applications. Area 38(2):165–174. https://doi.org/10.1111/j.1475-4762.2006.00671.x
6. Cavalli M, Trevisani S, Comiti F, Marchi L (2013) Geomorphometric assessment of spatial sediment connectivity in small Alpine catchments. Geomorphology 188:31–41. https://doi.org/10.1016/j.geomorph.2012.05.007
7. Chen C, Zhao G, Zhang Y, Bai Y, Tian P, Mu X, Tian X (2023) Linkages between soil erosion and long-term changes of landscape pattern in a small watershed on the Chinese Loess Plateau. CATENA 220(PA):106659. https://doi.org/10.1016/j.catena.2022.106659
8. Crema S, Cavalli M (2018) SedInConnect: a stand-alone, free and open source tool for the assessment of sediment connectivity. Comput Geosci 111:39–45. https://doi.org/10.1016/j.cageo.2017.10.009
9. Fargas D, Martínez Casasnovas JA, Poch R (1997) Identification of critical sediment source areas at regional level. Phys Chem Earth 22(3–4):355–359. https://doi.org/10.1016/S0079-1946(97)00158-4
10. Ferro V, Porto P (2000) Sediment delivery distributed (SEDD) model. J Hydrol Eng 5(4):411–422. https://doi.org/10.1061/(ASCE)1084-0699(2000)5:4(411)
11. Franz C, Makeschin F, Weiß H, Lorz C (2014) Sediments in urban river basins: Identification of sediment sources within the Lago Paranoá catchment, Brasilia DF, Brazil—using the fingerprint approach. Sci Total Environ 466–467:513–523. https://doi.org/10.1016/j.scitotenv.2013.07.056
12. Fryirs KA, Brierley GJ (2012) Geomorphic analysis of river systems. In: Geomorphic analysis of river systems. Wiley. https://doi.org/10.1002/9781118305454
13. Fryirs KA (2013) (Dis)Connectivity in catchment sediment cascades: a fresh look at the sediment delivery problem. Earth Surf Proc Land 38(1):30–46. https://doi.org/10.1002/esp.3242
14. Fryirs KA, Brierley GJ, Preston NJ, Spencer J (2007) Catchment-scale (dis)connectivity in sediment flux in the upper Hunter catchment, New South Wales, Australia. Geomorphology 84(3–4):297–316. https://doi.org/10.1016/j.geomorph.2006.01.044
15. Gay A, Cerdan O, Mardhel V, Desmet M (2016) Application of an index of sediment connectivity in a lowland area. J Soils Sediments 16(1):280–293. https://doi.org/10.1007/s11368-015-1235-y
16. Goldin B, Rudaz B, Bardou E (2016) Application of a sediment connectivity GIS-based index in a basin undergoing glacier retreat: the case study of the Navizence catchment. Rendiconti Online Della Società Geologica Italiana 39:35–38. https://doi.org/10.3301/ROL.2016.41
17. Guo L, Liu R, Men C, Wang Q, Miao Y, Shoaib M, Wang Y, Jiao L, Zhang Y (2021) Multiscale spatiotemporal characteristics of landscape patterns, hotspots, and influencing factors for soil erosion. Sci Total Environ 779:146474. https://doi.org/10.1016/j.scitotenv.2021.146474
18. Harvey AM (2001) Coupling between hillslopes and channels in upland fluvial systems: implications for landscape sensitivity, illustrated from the Howgill Fells, northwest England. CATENA 42(2–4):225–250. https://doi.org/10.1016/S0341-8162(00)00139-9

19. Hooke J (2003) Coarse sediment connectivity in river channel systems: a conceptual framework and methodology. Geomorphology 56(1–2):79–94. https://doi.org/10.1016/S0169-555X(03)00047-3
20. Greenfield J, Lahlou M, Swift L, Manguerra HB (2001) Watershed erosion and sediment load estimation tool. Soil Erosion. https://doi.org/10.13031/2013.3266
21. Jing Y, Zhao Q, Lu M, Wang A, Yu J, Liu Y, Ding S (2022) Effects of road and river networks on sediment connectivity in mountainous watersheds. Sci Total Environ 826:154189. https://doi.org/10.1016/j.scitotenv.2022.154189
22. Lemma H, Frankl A, van Griensven A, Poesen J, Adgo E, Nyssen J (2019) Identifying erosion hotspots in Lake Tana Basin from a multisite soil and water assessment tool validation: opportunity for land managers. Land Degrad Dev 30(12):1449–1467. https://doi.org/10.1002/ldr.3332
23. López-Vicente M, Nadal-Romero E, Cammeraat ELH (2017) Hydrological connectivity does change over 70 years of abandonment and afforestation in the Spanish Pyrenees. Land Degrad Dev 28(4):1298–1310. https://doi.org/10.1002/ldr.2531
24. Lu H, Moran CJ, Prosser IP (2006) Modelling sediment delivery ratio over the Murray Darling Basin. Environ Model Softw 21(9):1297–1308. https://doi.org/10.1016/j.envsoft.2005.04.021
25. Martini L, Cavalli M, Picco L (2022) Predicting sediment connectivity in a mountain basin: a quantitative analysis of the index of connectivity. Earth Surf Proc Land 47(6):1500–1513. https://doi.org/10.1002/esp.5331
26. Mishra K, Sinha R, Jain V, Nepal S, Uddin K (2019) Towards the assessment of sediment connectivity in a large Himalayan river basin. Sci Total Environ 661:251–265. https://doi.org/10.1016/j.scitotenv.2019.01.118
27. Najafi S, Sadeghi SH, Heckmann T (2021) Analysis of sediment accessibility and availability concepts based on sediment connectivity throughout a watershed. Land Degrad Dev 32(10):3023–3044. https://doi.org/10.1002/ldr.3964
28. Nelson EJ, Booth DB (2002) Sediment sources in an urbanizing, mixed land-use watershed. J Hydrol 264(1–4):51–68. https://doi.org/10.1016/S0022-1694(02)00059-8
29. Ortíz-Rodríguez AJ, Borselli L, Sarocchi D (2017) Flow connectivity in active volcanic areas: Use of index of connectivity in the assessment of lateral flow contribution to main streams. CATENA 157(January):90–111. https://doi.org/10.1016/j.catena.2017.05.009
30. Persichillo MG, Bordoni M, Cavalli M, Crema S, Meisina C (2018) The role of human activities on sediment connectivity of shallow landslides. Catena 160(September 2017):261–274. https://doi.org/10.1016/j.catena.2017.09.025
31. Poeppl RE, Keesstra SD, Maroulis J (2017) A conceptual connectivity framework for understanding geomorphic change in human-impacted fluvial systems. Geomorphology 277:237–250. https://doi.org/10.1016/j.geomorph.2016.07.033
32. Tamene L, Adimassu Z, Ellison J, Yaekob T, Woldearegay K, Mekonnen K, Thorne P, Le QB (2017) Mapping soil erosion hotspots and assessing the potential impacts of land management practices in the highlands of Ethiopia. Geomorphology 292:153–163. https://doi.org/10.1016/j.geomorph.2017.04.038
33. Wang A, Zhao Q, Yu Z, Yu J, Liu Y, Wang P, Zhang G, Zhou J, Ding S (2023) Factors and thresholds determining sediment delivery pathways between forest road and stream in mountainous watershed. Catena 224(November 2022):106976. https://doi.org/10.1016/j.catena.2023.106976

Chapter 6
Source-to-Sink Sediment Transfer Within the Catchments of Major Dams

Abstract This chapter investigates source-to-sink sediment connectivity within the major dams of the Damodar River Basin, focusing on the temporal dynamics of sediment transport and connectivity to sediment hotspots. The research examines how sediment moves from its source in upland erosion-prone areas to deposition zones within the dams and how this connectivity changes over time due to varying environmental factors and land use practices. A key aspect of this study is the Basin Dis-connectivity Index (BDI), which quantifies the degree of sediment transport interruption across the basin. The BDI is used to identify regions where sediment flow is hindered, leading to sediment accumulation or loss in specific areas, particularly in relation to sediment hotspots. The findings reveal how changes in land use, topography, and hydrological factors influence sediment connectivity and the formation of hotspots, contributing to long-term sedimentation patterns in the dams. This chapter highlights the importance of understanding source-to-sink connectivity for improving sediment management and mitigating sedimentation issues in the Damodar River Basin. It provides valuable insights for optimizing watershed management, reducing sedimentation risks, and enhancing the efficiency of the basin's major dams.

Keywords Sediment connectivity Index · Source-to-sink sediment connectivity · Basin disconnectivity · Soil erosion hotspot · Damodar River basin · Maithon dam · Panchet dam · Tenughat dam

6.1 Introduction

The construction of a dam within a river system alters the natural flow and sediment dynamics, reshaping processes and triggering long-term geomorphological and ecological changes throughout the basin or catchment area. The Damodar River Basin is managed by several large dams—Maithon, Panchet, Tenughat, Tilaya, and Konar—to mitigate the devastating effects of floods [3, 34]. Extensive mining and human activities in the basin's upper catchment have led to a substantial flux of

S. A. Alam and R. Maiti, *Sediment Source to Sink: Deciphering Sediment Connectivity to Large Dams in Damodar River Basin*, Earth and Environmental Sciences Library,
https://doi.org/10.1007/978-3-031-87072-9_6

eroded material into the river. This huge sediment load impacts downstream dams and reservoirs, predominantly reducing reservoir capacity and harming water quality. Therefore, understanding the natural flow and sediment transfer processes from the upper catchment of a river or dam system is essential for reservoir management and for mitigating sedimentation issues to maintain water storage capacity and dam functionality. Thus, the sediment connectivity plays a crucial role in understanding the linkage between sediment sources in the upper catchment and the sink [2, 4, 6–8, 13, 14, 19, 24, 29, 30, 41]. Consequently, A source-to-sink sediment connectivity model has been developed [4, 7] to trace sediment pathways from major sediment source zones to the reservoirs in dam catchments. Additionally, we can comprehend the influx of sediment load to the main channel and dams by quantifying sediment connectivity and (dis)connectivity.

6.2 Concepts of Source-to-Sink Relationship

Catchment or landscape connectivity refers to the transfer of flow, energy, and matter within and across ecosystems or fluvial systems [13, 40]. Numerous geomorphological studies have emphasized that sediment transfer from the source to the channel network is governed by primary processes that significantly influence landscape evolution. These processes determine the movement of sediments through fluvial systems, shaping landforms and impacting overall geomorphological development [18, 21]. In a particular fluvial system, the high mass transfer efficiency is also characterized by a significant degree of connectivity [5, 18, 21]. Various algorithms have been applied to estimate sediment delivery and sediment yield, addressing critical issues in sediment transport analysis. By integrating both upslope and downslope approaches, these methods provide a detailed understanding of sediment dynamics, enabling the identification of key processes and pathways influencing sediment movement [6, 7, 12, 21, 22, 27] for developing the index of connectivity (IC) approach. Therefore, understanding the source-to-sink relationship is crucial for watershed management, erosion control, and sedimentation analysis in reservoirs. It helps in predicting sediment yield, managing flood risks, and designing sustainable land-use practices in catchment areas. So, the nonlinear behaviors in the system can easily be identified and tracked through the source-to-sink connectivity approach (Fig. 6.2).

Connectivity is an emergent substance of a geomorphic system or fluvial system [40]. So, Connectivity can recognize several aspects of river processes and forms [39], including sediment transport, channel morphology, and ecosystem dynamics, by linking water, sediment, and biological elements throughout the river system. Therefore, various connectivity approaches have been developed to understand landscape ecology, hydrology, geomorphology, and hillslope-channel coupling as outcomes of system properties. A conceptual framework for sediment connectivity and hypothesis development was presented by [21] and, after, the conceptual framework was developed by [6], which incorporates various types of landscape linkages across different spatial scales within the catchment area. Landscape connectivity is a

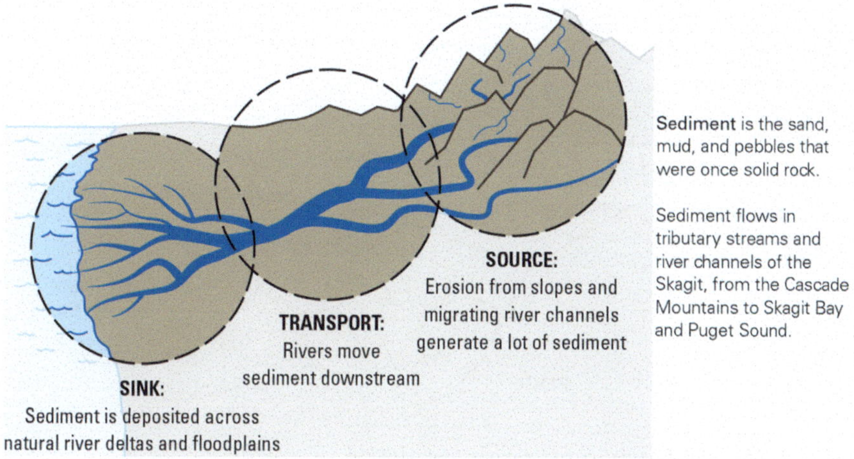

Fig. 6.1 Conceptual framework of source-to-sink connectivity in the river system [38]

primary control of catchment-scale fluxes of water and sediment [13]. The conceptual model for the sediment (dis)connectivity framework was meticulously developed by integrating longitudinal, lateral, and vertical blockages [14], providing a comprehensive approach to understanding multi-dimensional sediment flow impediments. The source-to-sink relationship explains the flux of energy and materials within a catchment or system through frameworks like sediment connectivity, hydrological connectivity, functional connectivity, landscape connectivity, and lateral and longitudinal connectivity. To understand the complexities of catchment systems, researchers worldwide have developed various connectivity approaches [4–7, 9–11, 13–15, 19, 20, 25, 26, 28, 29, 31–33, 35, 37, 39, 42, 44–46]. The GIS-based approach known as the Index of Connectivity (IC) was developed by [7, 10] to quantify sediment connectivity within a catchment, integrating spatial data and connectivity factors to assess how sediments are transferred across landscapes (Fig. 6.1).

6.3 Sediment Yield in the Major Dam Catchments

Sediment yield (SY) refers to the amount of sediment transported from a catchment area to a reservoir outlet over time. The concentration of sediment yield is influenced by geomorphological, hydrological, and human factors [23]. This chapter presents the simulation of monthly sediment yield for the outlets of the three major dam catchments—Maithon, Panchet, and Tenughat—in the Damodar Basin (Fig. 6.3).

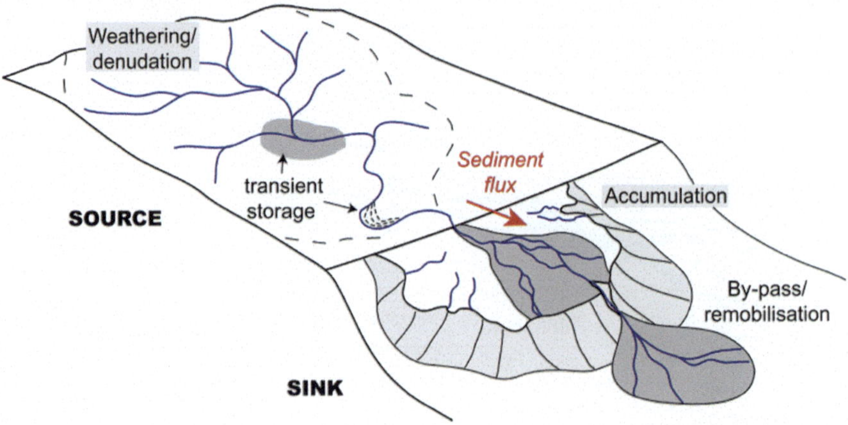

Fig. 6.2 Conceptual framework of source-to-sink [16]

6.3.1 Maithon Dam Catchment

The Maithon Dam catchment outlet has shown fluctuating sediment yield levels over the years, driven by factors such as land use, rainfall patterns, and human activity. The simulated sediment yield (SY) data from 1981 to 2013 shows considerable variability, with annual values ranging from a low of 10.96 t $ha^{-1}yr^{-1}$ in 1982 to a peak of 105.57 t $ha^{-1}yr^{-1}$ in 2011. The monthly sediment yield at the Maithon Dam outlet ranges from 0 to 57.24 t ha^{-1} (Fig. 6.4). The average annual sediment yield at the Maithon Dam outlet is approximately 52.48 t $ha^{-1}yr^{-1}$. The boxplot (Fig. 6.4) presents a detailed view of the sediment yield distribution, whereas the July month sediment yield distribution is characterized by a positively skewed pattern. Most data values are clustered at the lower end of the scale, signifying that low sediment yields are more frequent during this period. However, the plot exhibits a tail extending toward higher sediment yield values, indicating the presence of sporadic high-yield events. These events may correspond to extreme climatic conditions, such as heavy rainfall or flash floods, which are known to significantly influence sediment transport dynamics. Notably, the red dots visible in the plot represent statistical outliers, which are values that deviate markedly from the overall data distribution. These outliers likely signify extreme rainfall events within the Maithon dam catchments that produced huge amounts of sediment. Such events are critical as they not only highlight the impact of episodic weather conditions on sediment yield but also suggest underlying vulnerabilities in the catchment's erosion and sediment transport processes. Furthermore, a significant upward trend in sediment yield is observed ($R^2 = 0.193$, $p < 0.01$). Increasing sediment yields can reduce reservoir storage capacity over time, deteriorate water quality, and impact downstream ecosystems.

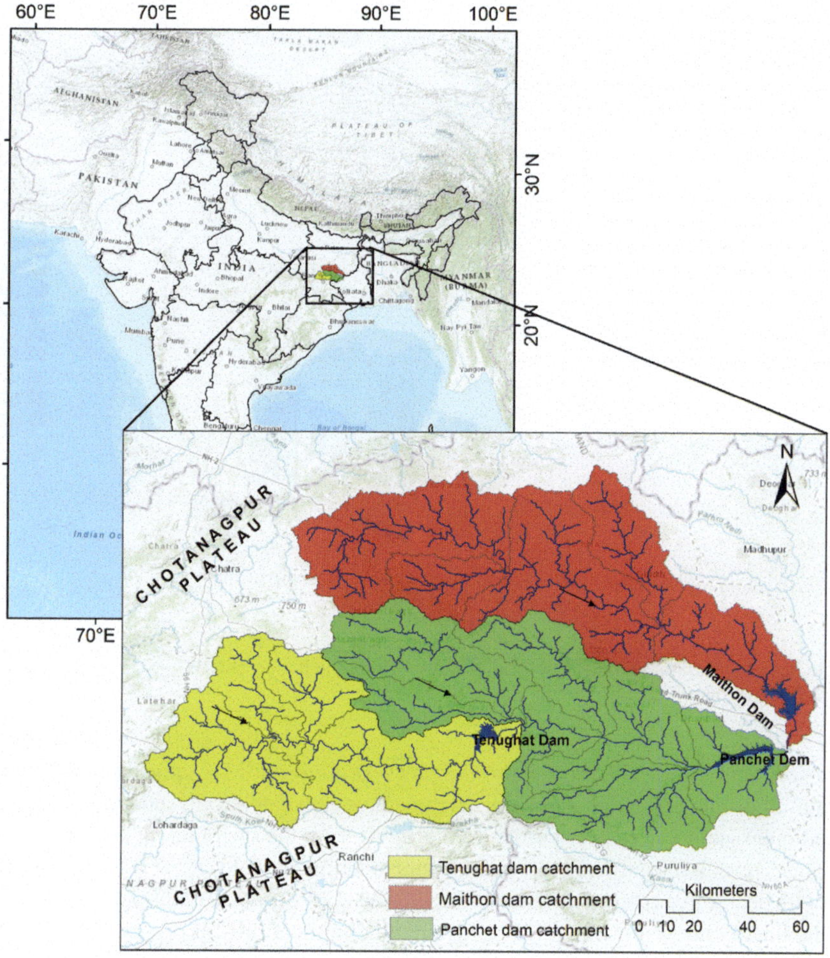

Fig. 6.3 Location map depicting the three major dams in the Damodar River Basin (DRB)

6.3.2 Panchet Dam Catchment

The sediment yield at the Panchet Dam catchment outlet has exhibited variability over the years. The simulated sediment yield (SY) data from 1981 to 2013 indicate significant changes over the observed period, ranging from a low of 16.02 t $ha^{-1}yr^{-1}$ in 1982 to a maximum of 89 t $ha^{-1}yr^{-1}$ in 2013. The monthly sediment yield at the Panchet Dam outlet ranges from 0 to 47.24 t ha^{-1} (Fig. 6.4). The average annual sediment yield at the Panchet Dam outlet is approximately 38.39 t $ha^{-1}yr^{-1}$. The boxplot (Fig. 6.5) for June and October at the Panchet Dam catchment outlet shows a positively skewed distribution of sediment yield, indicating that most sediment yield values are relatively low, while infrequent extreme events cause significant peaks

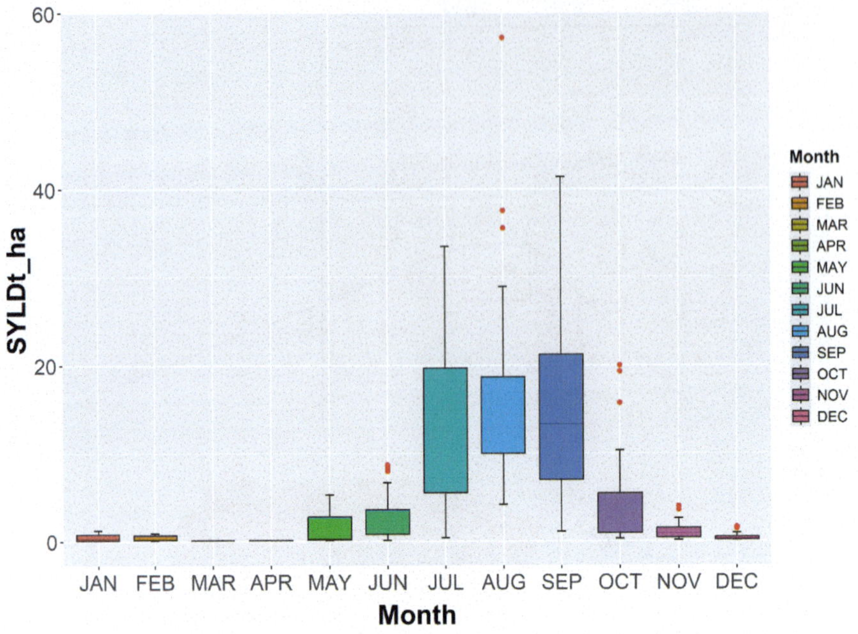

Fig. 6.4 Mean monthly sediment yield variation within the period of (1981–2013) in the Maithon dam catchment

in sediment yield. The red dots in the plot represent outliers, indicating significant rainfall events in the Panchet Dam catchment, which can lead to sudden increases in sediment yield as a result of erosion and runoff from the surrounding landscape. The dataset reveals a significant increasing trend in sediment yield ($R^2 = 0.25, p < 0.01$), ,, with peak values recorded in 2011 (73.493 t $ha^{-1}yr^{-1}$) and 2012 (69.347 t $ha^{-1}yr^{-1}$). The fluctuating sediment yields have potential implications for reservoir sedimentation and water quality, as high sediment loads can reduce the storage capacity of the dam, impact water treatment processes, and degrade water quality.

6.3.3 Tenughat Dam Catchment

The Panchet Dam catchment outlet has shown fluctuating sediment yield levels over the years. The simulated sediment yield (SY) data from 1981 to 2013 show significant changes in sediment yield, ranging from a low of 16.19 t $ha^{-1}yr^{-1}$ in 1983 to a peak of 115.65 t $ha^{-1}yr^{-1}$ in 2008. The estimated sediment yield at the Panchet Dam outlet ranges from 0 to 67.91 t ha^{-1} (Fig. 6.6). The average annual sediment yield at the Panchet Dam outlet is approximately 63.76 t $ha^{-1}yr^{-1}$. The sediment yield shows a significant increasing trend ($R^2 = 0.16, p < 0.01$), with peak values observed in 2004 (91.57 t $ha^{-1}yr^{-1}$), 2008 (115.65 t $ha^{-1}yr^{-1}$), and

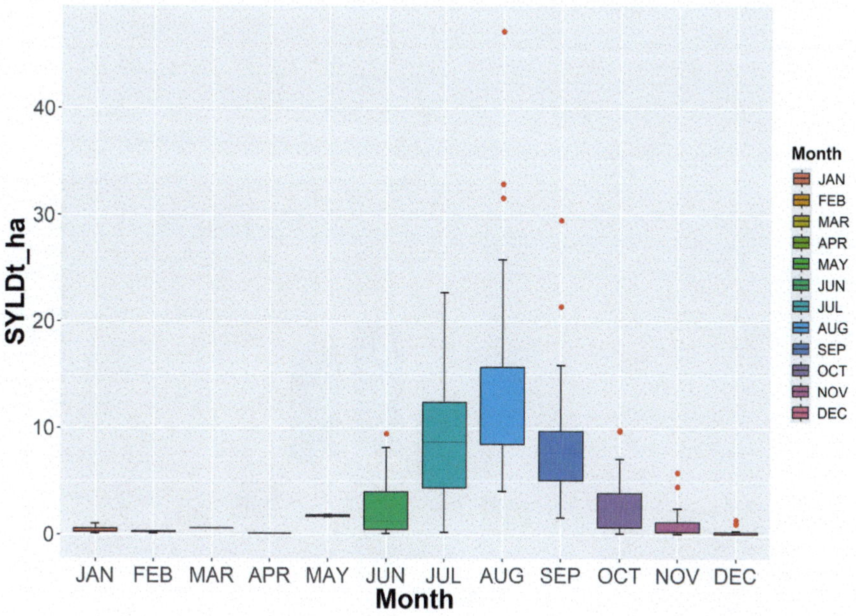

Fig. 6.5 Mean monthly sediment yield variation within the period of (1981–2013) in the Panchet dam catchment

1997 (111.99 t $ha^{-1}yr^{-1}$). Conversely, years such as 1995 (20.67 t $ha^{-1}yr^{-1}$) and 1983 (16.19 t $ha^{-1}yr^{-1}$) exhibit much lower sediment yields. The boxplot for June, July, August, and September display a positively skewed distribution of sediment yield, indicating that sediment yield values are clustered at lower production, with a tail extending toward higher sediment yield (Fig. 6.6). The red dots in the plot represent outliers indicating high sediment yield which is linked to extreme rainfall events in the Tenughat Dam catchment. Intense rainfall leads to high surface runoff and soil erosion, mobilizing large amounts of sediment. These extreme events, when coupled with human-induced factors such as open-cast mining and deforestation, significantly exacerbate sediment yield.

6.4 Source-to-Sink Sediment Connectivity Over Time

The dynamic assessment of sediment routes is essential for leveraging a thorough understanding of sediment source and sink areas and their transfer mechanism [17]. The production of sediment yield in the outlet is determined by the effective integration of the sediment source zone and the sediment connectivity (Fig. 6.7) or (dis)connectivity. Sediment connectivity reflects the accessibility of sediment within

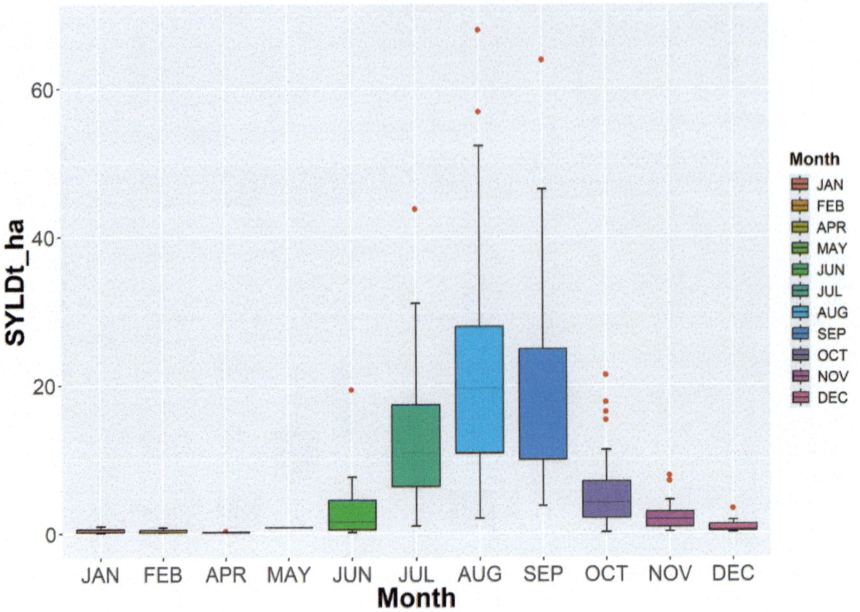

Fig. 6.6 Mean monthly sediment yield variation within the period of (1981–2013) in Tenughat Dam

a system, while soil erosion represents the quantity of sediment available in the watershed [29]. Moreover, the integration of sediment connectivity with soil erosion highlights the intensity of sediment input to the channel and the resulting sediment yield at the catchment outlet. Sediment transport dynamics have evolved due to natural factors and anthropogenic interventions like land use changes, deforestation, and/or dam construction. In this section, we discussed source-to-sink sediment connectivity over time in the three major dam catchments–Maithon, Panchet, and Tenughat dam catchments.

6.4.1 Maithon Dam Catchment

The sediment connectivity values for the Maithon dam catchment from 1990 to 2020 reveal critical trends in sediment dynamics. The minimum values demonstrate a slight rise in negative connectivity over time, rising from − 4.57 in 1990 to − 5.45 in 2005, indicating persistent areas of sediment dis-connectivity. Conversely, the positive connectivity values exhibit a consistent upward trend, increasing from 3.37 in 1990 to 3.62 in 2020. The mean values remain relatively stable, fluctuating around − 1.37 to − 1.40, suggesting a consistent level of connectivity within the catchment. Additionally, the standard deviation values remain consistent (0.77–0.79), reflecting a uniform variability in connectivity across the years. The high

Fig. 6.7 The field photograph provides visual evidence that supports the concept of efficient sediment transfer and connectivity within the catchment (Photo credit: Sk Asraful Alam)

sediment connectivity was observed in the Parasnath hill region, Jamkol, Tunda, Kutma, and Lemua. Overall, these trends underscore a complex interplay of sediment connectivity dynamics in the Maithon catchment.

6.4.2 Panchet Dam Catchment

The sediment connectivity values for the Panchet dam catchment over the years 1990, 2005, and 2020 reveal a stable yet negative trend in sediment connectivity, with mean values consistently around − 1.26. The minimum connectivity values exhibited a trend toward greater negativity, particularly from − 4.66 in 1990 to − 5.56 in 2005, suggesting enhanced sediment retention in the catchment. The positive connectivity values showed little variation, peaking at 3.42 in 1990 and 3.45 in 2020, suggesting stable positive connectivity during peak flow events. The consistent standard deviation around 0.82 indicates minimal variability in these connectivity assessments, suggesting that the underlying geomorphological characteristics of the catchment have remained stable over time. Significant sediment connectivity was noted in the Parasnath Hill region, Jumra Pahar, and the Chatra area. Overall, there is positive sediment connectivity in the hilly terrain characterized by moderate slopes.

6.4.3 *Tenughat Dam Catchment*

The sediment connectivity values for the Tenughat dam catchment across the years 1990, 2005, and 2020 reveal significant trends in sediment dynamics. The negative connectivity values reduced slightly from − 4.37 in 1990 to − 4.55 in 2020, indicating a consistent low connectivity environment. The positive connectivity values remained relatively stable, peaking at 3.81 in 1990 to 3.80 in 2020. This shift reflects changing hydrological or geomorphological conditions that may have enhanced sediment transport processes within the catchment. The standard deviation values remained consistent (ranging from 0.847 to 0.86), indicating relatively stable variability in sediment connectivity across the observed years. Significant sediment connectivity was noted in the Ramgarh Hill region, Patratu region, Tandwa hilly area, Chatra region, Jumra Pahar, and Lugu Pahar. This evaluation underscores the importance of monitoring to understand sediment dynamics in response to environmental changes and land use practices.

6.5 Connectivity to Soil Erosion Hotspot (SEH)

In this section, we intend to elucidate the sediment connectivity pathways from the significant ($p < 0.01$) soil erosion areas to the sink reservoirs or primary channels in the three major dams (Maithon, Panchet, and Tenughat) catchment. Soil erosion hotspots are areas within a catchment that contribute a significant amount of sediment load in the channel. Sediment connectivity is crucial for understanding sediment transfer mechanisms from various sediment sources to primary channels [4, 7, 20]. Understanding sediment connectivity to sediment hotspots is essential for effective watershed management and for forecasting the implications of sedimentation in reservoirs within these dam catchments. Therefore, The linkage between sediment connectivity and erosional [1, 8] hotspots is pivotal for understanding sediment dynamics in dam catchments.

6.5.1 *Maithon Dam Catchment*

The connectivity pathways from erosional hotspots to the sink were investigated in the Maithon Dam catchment for the years 1990, 2005, and 2020 (Fig. 6.8). This analysis considers regions where sediment connectivity exceeds an IC threshold of 0.5 (IC > 0.5), with soil erosion displaying high statistical significance ($p < 0.01$). Findings indicate that significant erosional areas with improved sediment connectivity demonstrate a greater likelihood of sediment transport to the sink (Reservoir). Consequently, erosional hotspots with robust connectivity pathways are most likely to contribute substantial sediment volumes to the main channel. In 1990, there were

18 identified erosional hotspots covering an area of 449.60 km^2, whereas, by 2020, the number of hotspots reduced to 16, covering a slightly larger area of 453.97 km^2 (Fig. 6.8). In 1990, the largest erosional hotspot covered an area of 67.96 km^2; by 2020, a single hotspot had expanded to 216.28 km^2. The number of hotspots has decreased over time, although their area coverage is expanding in this region. In 1990, all erosional hotspots showed high sediment connectivity (IC) values ranging from 1.6 to 2.97, whereas in 2020, the IC values ranged from 1.52 to 3.42, indicating a high coupling with valley-side terrain with the stream channel, which facilitates easy transfer of sediment from the surrounding region to the channel and ultimately in the dam. The combination of erosional hotspots and high sediment connectivity in this region suggests a greater likelihood of maximum sediment production in the main channel. This indicates that these regions contributed the highest sediment input to both the main channel and the sink (reservoir).

6.5.2 *Panchet Dam Catchment*

The sediment connectivity pathways from erosional hotspots to the sink were assessed in the Tenughat Dam catchment for the years 1990, 2005, and 2020 (Fig. 6.9). The analysis focused on areas where sediment connectivity surpassed an IC threshold of 0.5 (IC > 0.5), with soil erosion showing a high degree of statistical significance ($p < 0.01$). The results suggest that areas with significant erosion and improved connectivity are more likely to transport sediment to the sink (Reservoir). Erosional hotspots exhibiting strong connectivity pathways are, therefore, more likely to contribute substantial sediment to the main channel. In 1990, there were 26 erosional hotspots covering 669.70 km^2, while by 2020, this number reduced to 15, but the area covered by the hotspots slightly increased to 705.934 km^2 (Fig. 6.9). The primary reason for the increased areal coverage of hotspots is the spatial expansion and merging with neighboring hotspots. In 1990, the largest erosional hotspot spanned 96.01 km^2, but by 2020, one hotspot had expanded to 238.75 km^2. Despite the decrease in the number of hotspots over time, the total area of these hotspots has continued to grow. In 1990, all hotspots had high sediment connectivity (IC) values ranging from 1.6 to 3.42, whereas by 2020, these values ranged from 1.52 to 3.44. The combination of erosional hotspots with high sediment connectivity suggests an increased potential for maximum sediment production in the main channel. This signifies that these areas have contributed the largest sediment input to both the main channel and the sink (reservoir).

6.5.3 *Tenughat Dam Catchment*

The connectivity between erosional hotspots and the sediment sink in the Tenughat Dam catchment was examined for the years 1990, 2005, and 2020 (Fig. 6.10). This

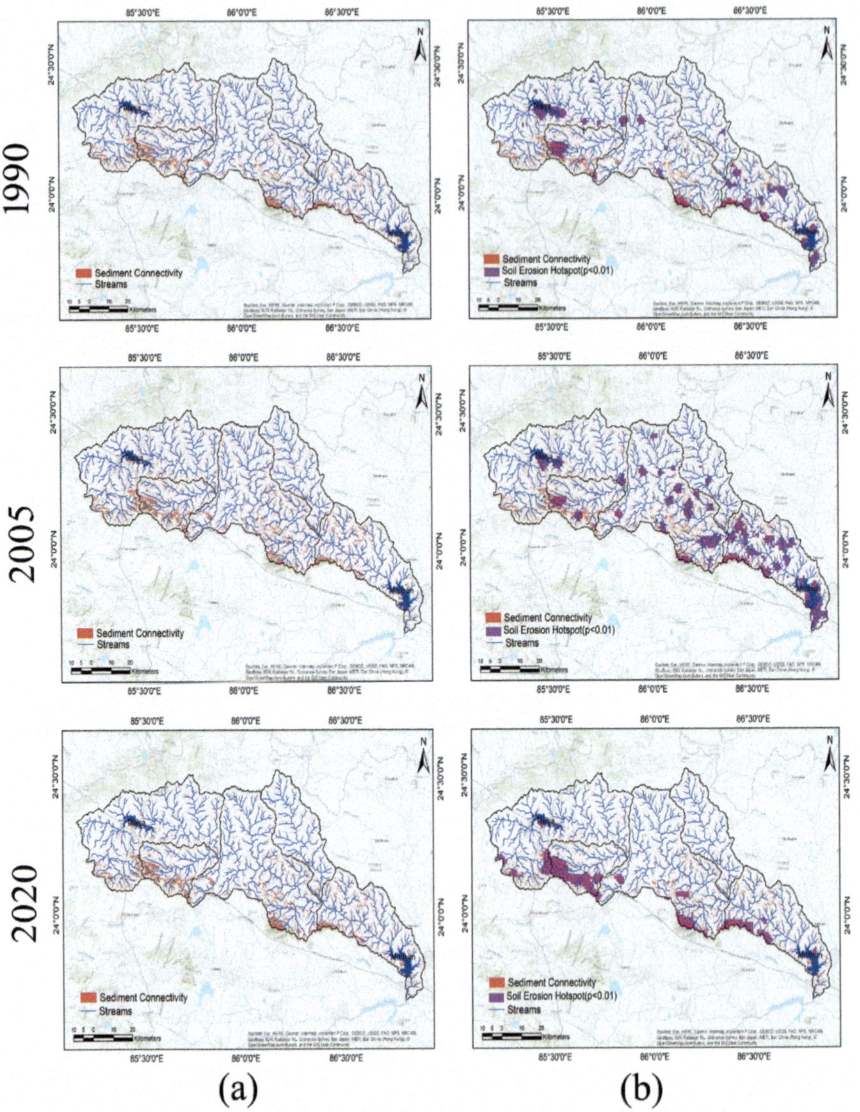

Fig. 6.8 Sediment connectivity, (a) significant sediment connectivity regions, (b) significant soil erosion hotspots (p< 0.01) reside in the highest sediment connectivity regions in the Maithon catchment

study focused on areas where sediment connectivity exceeded the threshold of 0.5 (IC > 0.5), with soil erosion showing statistically significant results ($p < 0.01$). The findings highlight that regions with higher erosion and better connectivity are more likely to facilitate sediment transport toward the sink (reservoir). Consequently, hotspots with enhanced connectivity pathways are expected to contribute large amounts of

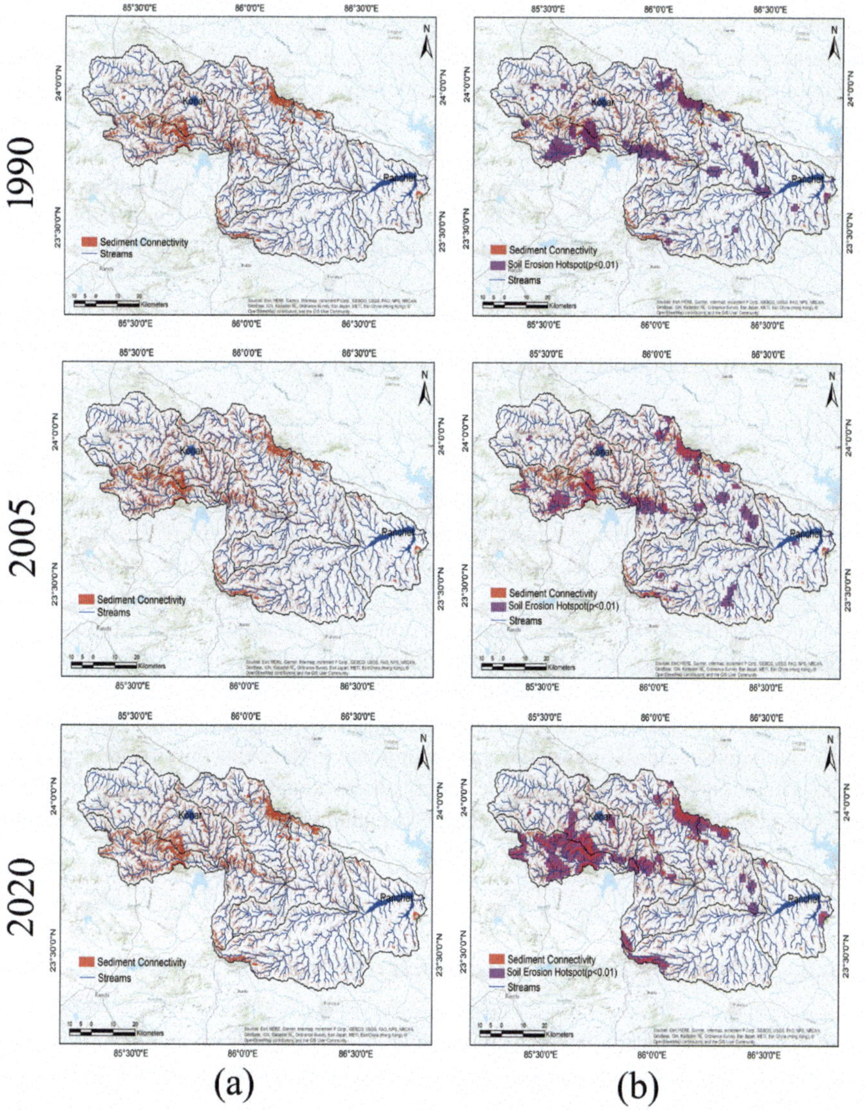

Fig. 6.9 Sediment connectivity, (a) significant sediment connectivity regions, (b) significant soil erosion hotspots ($p< 0.01$) reside in the highest sediment connectivity regions in the Panchet Catchment

sediment to the main channel. In 1990, there were 25 erosional hotspots, covering 1077.19 km^2, but by 2020, the number decreased to 13, despite a slight increase in the total area to 1264.93 km^2 (Fig. 6.10). The primary reason for the increased areal coverage of hotspots is the spatial expansion and merging with neighboring hotspots. The largest erosional hotspot in 1990 occupied 162.48 km^2, while by 2020, the largest hotspot had expanded to 467.84 km^2. The growth of these hotspots has been driven by anthropogenic activities such as open-cast mining, deforestation, and stone mining. Although the total number of hotspots has declined, their area has increased significantly due to these human impacts. In 1990, IC values for all hotspots ranged from 0.15 to 3.31, while in 2020, the IC values spanned from 1.59 to 3.81. The presence of large erosional hotspots with high sediment connectivity (IC > 1.59) suggests that these areas are key contributors to maximum sediment production, with substantial sediment flowing into the main channel and ultimately to the sink (reservoir).

6.6 Basin Dis-connectivity Index (BDI)

The Basin Disconnectivity Index (BDI) is a measure used to assess the level of connectivity or disconnectivity within a river basin or watershed, based on a river classification system [43]. This classification system (Fig. 6.11) serves as the basis for calculating BDI, providing insights into the connectivity status of the river basin. The classification framework proposed by [36], reflects the extent to which various landscape and hydrological elements, such as terrain, vegetation, dams, and anthropogenic barriers, affect the movement and continuity of water, sediment, and other materials across the basin. The BDI can indicate how effectively a basin's structure allows or hinders sediment transfer from upstream to downstream areas, resulting in an important tool for evaluating sediment transport dynamics, erosion vulnerability, and watershed health. Higher BDI values indicate greater disconnection, implying less sediment reaching the dam, while lower values indicate higher connectivity. This section may discuss BDI's role in watershed management, sediment retention, and implications for reservoir sedimentation.

6.6.1 *Concepts of Basin Dis-connectivity*

The concept of the basin disconnectivity index (BDI) focuses on understanding and quantifying how sediment transport pathways are interrupted or disconnected within a river basin. In this study, the BDI was calculated [36, 43] to assess the extent of disconnection within the Damodar River basin resulting from dam development and river regulation. The BDI was calculated using Eq. (6.1).

$$BDI = \frac{\sum_{i=1}^{n} (ns_i^2 - ns_i) \times wp_i}{NS^2 - NS} \times 100\% \tag{6.1}$$

Here,

BDI Basin Dis-connectivity Index
ns_i Total number of distinct river systems related to channel network i

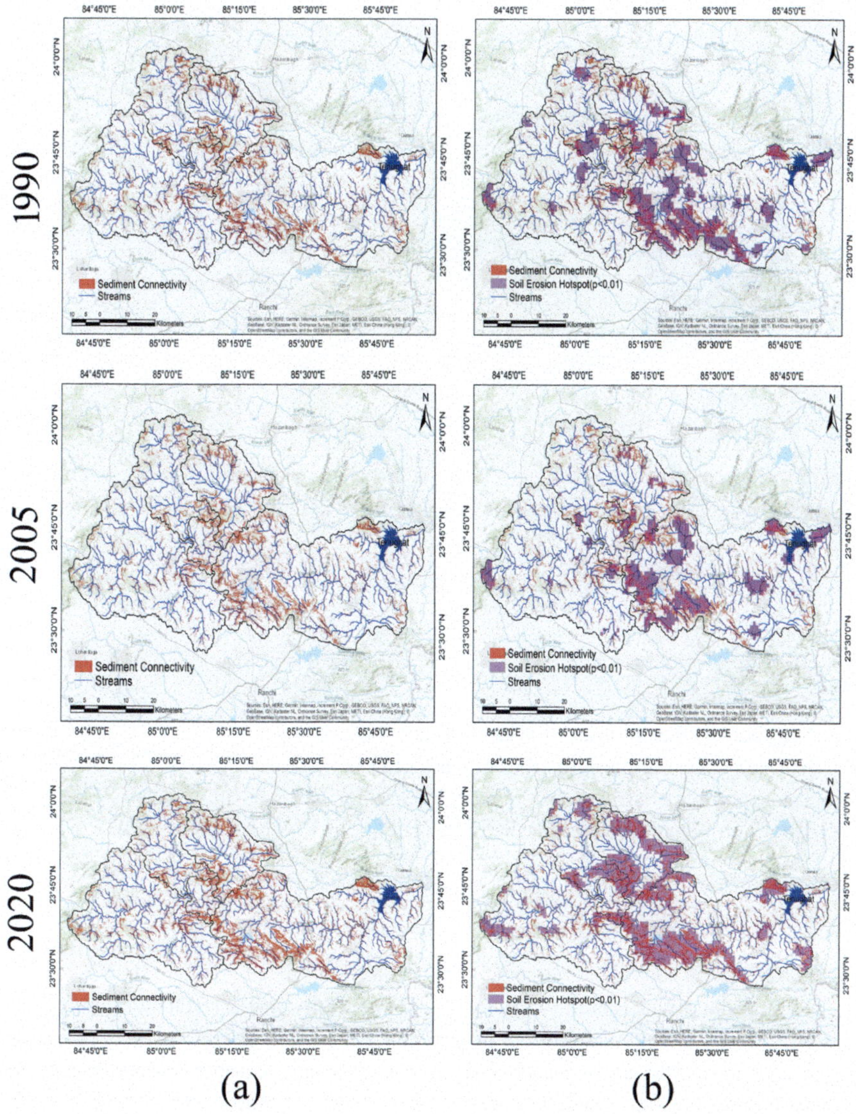

Fig. 6.10 Sediment connectivity, **a** significant sediment connectivity regions, **b** Significant soil erosion hotspots ($p < 0.01$) reside in the highest sediment connectivity regions in the Tenughat Catchment

NS Total number of river systems found in the Damodar River basin

wp_i The weighted percentage of channel length in section i relative to the total channel network length, which is defined as:

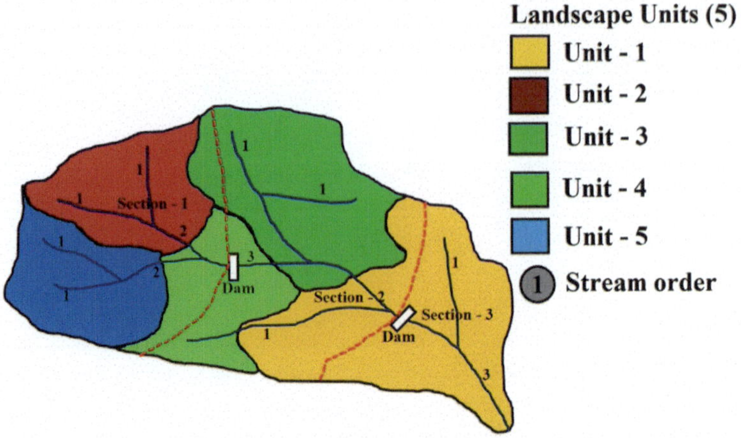

Fig. 6.11 Conceptual diagram illustrating the basin Disconnectivity index (BDI) based on channel length and river-system classification map [43] for the Damodar River basin

$$wp_i = \frac{\sum_{j=1}^{i} (w_j \times l_j)}{\sum_{k=1}^{k} (w_k \times l_k)} \tag{6.2}$$

Here,

w_j Number of stream orders for each channel in section j
l_j Number of stream lengths for each channel in section j
w_k Total number of stream orders in the Damodar River basin
l_k Total number of stream lengths in the Damodar River basin.

6.6.2 *Basin Dis-Connectivity Index in Damodar Basin*

In the Damodar River Basin, the Basin Disconnectivity Index (BDI) was computed to assess the efficiency of sediment transport across different sub-basins, evaluating the degree of connectivity or disconnectivity between sediment sources and the downstream reservoirs. The study identified five distinct landscape units (Fig. 6.10) following the proposed classification framework [36, 43]. The BDI value of the Damodar River basin is 66.67, indicating substantial disconnectivity primarily caused by dam structures. This high disconnectivity indicates that a significant portion of sediment generated from the upstream is retained by these reservoirs. By reducing sediment flow to downstream reservoirs, this process can assist retard sediment accumulation, consequently extending the reservoirs' operational life and conserving their storage capacity for a longer length of time. Key dams such as Tenughat, Panchet, and Maithon have a critical role in sediment trapping and control within the Damodar Basin. In contrast, reservoirs like Konar and Tilaiya exhibit lower sediment retention impacts than other structures.

Furthermore, the first-order streams contribute the longest combined stream length (e.g., 677.82 km for Panchet and 589.70 km for Maithon), which emphasizes the extensive network of initial sediment transport pathways. Second-order streams follow with 361.02 km (Panchet) and 284.66 km (Maithon). The stream order lengths reveal that lower-order streams (1st and 2nd) have extensive lengths across these basins (Table 6.1), reflecting more widespread sediment pathways in upper catchments, while higher-order stream segments contribute less significantly in length. This variation in stream order distribution shows that most sediment movement occurs in upper reaches, with first and second-order streams contributing the most to sediment flow. As stream order increases, the cumulative length decreases, showing that higher-order channels play a less extensive role in direct sediment transport, highlighting the role of smaller tributaries in upper catchments in sediment delivery. These metrics underscore the role of dams in reducing connectivity by interrupting sediment flow from lower to higher-order streams, establishing controlled sediment retention that aids in the long-term maintenance of reservoir capacities across the Damodar Basin.

6.7 Major Findings

The major findings of this chapter are as follows.

- The average annual sediment yield was 52.48 t $ha^{-1}yr^{-1}$ at Maithon dam, Whereas, 38.39 t $ha^{-1}yr^{-1}$ at Panchet Dam and 63.76 t $ha^{-1}yr^{-1}$ at Tenughat Dam.
- The highest sediment yield observed at Tenughat Dam is primarily driven by anthropogenic factors, including extensive open-cast mining, deforestation, and other human activities.
- Positive sediment connectivity was observed in the Piedmont and hilly regions, as well as near the main channel, and across all dam catchments, reflecting efficient sediment transport and connection between these areas and the river systems.
- The area coverage of hotspots across all catchments in the Damodar River basin from 1990 to 2020 has increased due to spatial expansion and the merging of adjacent hotspots.
- The BDI value of 66.67 for the Damodar River basin indicates significant disconnectivity, primarily caused by the presence of dam structures.

6.8 Conclusions

The study of source-to-sink sediment transfer within the catchments of major dams in the Damodar Basin provides comprehensive insights into sediment dynamics, connectivity, and its implications for watershed management. The findings reveal significant variations in sediment yield across the catchments, with the highest

Table 6.1 Description of Basin Dis-connectivity index (BDI) based on stream order length and landscape units

Total watershed		Name of the Dam						
		Konar	Tenughat	Panchet	Tilaiya	Maithon	Durgapur Barrage	Damodar
Stream order	Length (km)							
1st	2694.19	94.65	544.81	677.82	118.85	589.70	258.68	409.67
2nd	1333.23	76.17	203.07	361.02	53.94	284.66	165.93	188.42
3rd	598.80	48.17	148.86	176.82	48.82	122.69	36.07	17.35
4th	371.45		137.72	18.42		201.70	13.60	0
5th	342.58			103.46			73.22	165.89
wpi		0.03781	0.18815	0.24347	0.03604	0.22538	0.10809	0.16107
wpi (%)		3.78	18.81	24.35	3.60	22.54	10.81	16.11
Basin disconnectivity index (BDI)		66.67						

sediment yield at Tenughat, influenced by anthropogenic factors such as open-cast mining and deforestation. The analysis of sediment transport pathways indicates positive connectivity in the Piedmont, hilly regions, and near the main channel, highlighting efficient sediment transport. However, soil erosion hotspots have expanded over time, emphasizing the need for targeted erosion control measures. The analysis highlights critical soil erosion hotspots (SEH) that contribute substantially to sediment yield, emphasizing the need for targeted erosion control measures. The Basin Dis-connectivity Index (BDI) offers a valuable tool to quantify spatial and temporal disruptions in sediment flow, indicating areas prone to sediment trapping or loss. The Basin Dis-connectivity Index (BDI) of 66.67 suggests substantial dis-connectivity, primarily caused by dam structures that obstruct sediment flow. This research underscores the importance of understanding sediment connectivity for effective watershed management, highlighting the implications for dam sedimentation and the sustainability of water resources. Enhanced sediment management strategies, focusing on reducing dis-connectivity, are essential for maintaining dam reservoir capacity and mitigating erosion-related risks. The findings contribute to more informed decision-making regarding sediment source identification and its management within the Damodar Basin. Enhanced strategies focusing on reducing dis-connectivity are crucial for maintaining sediment flow, preventing sediment trapping, and ensuring the long-term sustainability of water resources in the Damodar Basin.

References

1. Abebe N, Eekhout J, Vermeulen B, Boix-Fayos C, de Vente J, Grum B, Hoitink T, Baartman J (2023) The potential and challenges of the 'RUSLE-IC-SDR' approach to identify sediment dynamics in a Mediterranean catchment. CATENA 233(100):107480. https://doi.org/10.1016/j.catena.2023.107480
2. Ardizzone F, Esposito G, Cavalli M, Crema S, Fiorucci F (2024) Sediment connectivity as a key to understand geomorphic effects of the Storm Alex in two mountain catchments of the Mediterranean Alps (Italy). Geomorphology 455(July 2023):109176. https://doi.org/10.1016/j.geomorph.2024.109176
3. Bhattacharyya K (2011) The lower Damodar River, India. Springer, Netherlands. https://doi.org/10.1007/978-94-007-0467-1
4. Borselli L, Cassi P, Torri D (2008) Prolegomena to sediment and flow connectivity in the landscape: a GIS and field numerical assessment. CATENA 75(3):268–277. https://doi.org/10.1016/j.catena.2008.07.006
5. Bracken LJ, Croke J (2007) The concept of hydrological connectivity and its contribution to understanding runoff-dominated geomorphic systems. Hydrol Process 21(13):1749–1763. https://doi.org/10.1002/hyp.6313
6. Brierley G, Fryirs K, Jain V (2006) Landscape connectivity: the geographic basis of geomorphic applications. Area 38(2):165–174. https://doi.org/10.1111/j.1475-4762.2006.00671.x
7. Cavalli M, Trevisani S, Comiti F, Marchi L (2013) Geomorphometric assessment of spatial sediment connectivity in small Alpine catchments. Geomorphology 188:31–41. https://doi.org/10.1016/j.geomorph.2012.05.007

8. Chen C, Zhao G, Zhang Y, Bai Y, Tian P, Mu X, Tian X (2023) Linkages between soil erosion and long-term changes of landscape pattern in a small watershed on the Chinese Loess Plateau. CATENA 220(PA):106659. https://doi.org/10.1016/j.catena.2022.106659
9. Covino T (2017) Hydrologic connectivity as a framework for understanding biogeochemical flux through watersheds and along fluvial networks. Geomorphology 277(September 2016):133–144. https://doi.org/10.1016/j.geomorph.2016.09.030
10. Crema S, Cavalli M (2018) SedInConnect: a stand-alone, free and open source tool for the assessment of sediment connectivity. Comput Geosci 111:39–45. https://doi.org/10.1016/j.cageo.2017.10.009
11. Croke J, Fryirs K, Thompson C (2013) Channel-floodplain connectivity during an extreme flood event: implications for sediment erosion, deposition, and delivery. Earth Surf Proc Land 38(12):1444–1456. https://doi.org/10.1002/esp.3430
12. Ferro V, Porto P (2000) Sediment delivery distributed (SEDD) model. J Hydrol Eng 5(4):411–422. https://doi.org/10.1061/(ASCE)1084-0699(2000)5:4(411)
13. Fryirs KA, Brierley GJ (2012) Geomorphic analysis of river systems. In: Geomorphic analysis of river systems. Wiley. https://doi.org/10.1002/9781118305454
14. Fryirs K (2013) (Dis)Connectivity in catchment sediment cascades: A fresh look at the sediment delivery problem. Earth Surf Proc Land 38(1):30–46. https://doi.org/10.1002/esp.3242
15. Gay A, Cerdan O, Mardhel V, Desmet M (2016) Application of an index of sediment connectivity in a lowland area. J Soils Sediments 16(1):280–293. https://doi.org/10.1007/s11368-015-1235-y
16. Grimaud J, Rouby D, Chardon D, Beauvais A (2018) Cenozoic sediment budget of West Africa and the Niger delta. Basin Res 30(2):169–186. https://doi.org/10.1111/bre.12248
17. Hao R, Huang X, Cai ZW, Xiao HB, Wang J, Shi ZH (2022) Incorporating sediment connectivity index into MUSLE model to explore soil erosion and sediment yield relationships at event scale. J Hydrol 614(PB):128579. https://doi.org/10.1016/j.jhydrol.2022.128579
18. Harvey AM (2001) Coupling between hillslopes and channels in upland fluvial systems: implications for landscape sensitivity, illustrated from the Howgill Fells, northwest England. CATENA 42(2–4):225–250. https://doi.org/10.1016/S0341-8162(00)00139-9
19. Heckmann T, Cavalli M, Cerdan O, Foerster S, Javaux M, Lode E, Smetanová A, Vericat D, Brardinoni F (2018) Indices of sediment connectivity: opportunities, challenges and limitations. Earth Sci Rev 187(August):77–108. https://doi.org/10.1016/j.earscirev.2018.08.004
20. Heckmann T, Schwanghart W (2013) Geomorphic coupling and sediment connectivity in an alpine catchment—exploring sediment cascades using graph theory. Geomorphology 182:89–103. https://doi.org/10.1016/j.geomorph.2012.10.033
21. Hooke J (2003) Coarse sediment connectivity in river channel systems: a conceptual framework and methodology. Geomorphology 56(1–2):79–94. https://doi.org/10.1016/S0169-555X(03)00047-3
22. Greenfield J, Lahlou M, Swift L Jr, Manguerra HB (2001) Watershed erosion and sediment load estimation tool. Soil Erosion. https://doi.org/10.13031/2013.3266
23. Kanito D, Bedadi B, Feyissa S (2023) Sediment yield estimation in GIS environment using RUSLE and SDR model in Southern Ethiopia. Geomat Nat Hazards Risk 14(1). https://doi.org/10.1080/19475705.2023.2167614
24. Koreňová S, Michalková MŠ, Máčka Z, Pöppl R, Donoval J (2024) Linking sediment connectivity with sediment transport risk assessment in small forested catchments in the Czech Republic. River Res Appl. https://doi.org/10.1002/rra.4295
25. Latocha A (2014) Geomorphic connectivity within abandoned small catchments (Stołowe Mts, SW Poland). Geomorphology 212:4–15. https://doi.org/10.1016/j.geomorph.2013.04.030
26. López-Vicente M, Ben-Salem N (2019) Computing structural and functional flow and sediment connectivity with a new aggregated index: a case study in a large Mediterranean catchment. Sci Total Environ 651:179–191. https://doi.org/10.1016/j.scitotenv.2018.09.170
27. Lu H, Moran CJ, Prosser IP (2006) Modelling sediment delivery ratio over the Murray Darling Basin. Environ Model Softw 21(9):1297–1308. https://doi.org/10.1016/j.envsoft.2005.04.021

28. Martini L, Cavalli M, Picco L (2022) Predicting sediment connectivity in a mountain basin: a quantitative analysis of the index of connectivity. Earth Surf Proc Land 47(6):1500–1513. https://doi.org/10.1002/esp.5331
29. Najafi S, Dragovich D, Heckmann T, Sadeghi SH (2021) Sediment connectivity concepts and approaches. Catena 196(March 2020):104880. https://doi.org/10.1016/j.catena.2020.104880
30. Najafi S, Sadeghi SH, Heckmann T (2021) Analysis of sediment accessibility and availability concepts based on sediment connectivity throughout a watershed. Land Degrad Dev 32(10):3023–3044. https://doi.org/10.1002/ldr.3964
31. Ortíz-Rodríguez AJ, Borselli L, Sarocchi D (2017) Flow connectivity in active volcanic areas: use of index of connectivity in the assessment of lateral flow contribution to main streams. CATENA 157(January):90–111. https://doi.org/10.1016/j.catena.2017.05.009
32. Poeppl RE, Fryirs KA, Tunnicliffe J, Brierley GJ (2020) Managing sediment (dis)connectivity in fluvial systems. Sci Total Environ 736. https://doi.org/10.1016/j.scitotenv.2020.139627
33. Poeppl RE, Keesstra SD, Maroulis J (2017) A conceptual connectivity framework for understanding geomorphic change in human-impacted fluvial systems. Geomorphology 277:237–250. https://doi.org/10.1016/j.geomorph.2016.07.033
34. Rudra K (2018) Rivers of the Ganga-Brahmaputra-Meghna Delta. Springer International Publishing. https://doi.org/10.1007/978-3-319-76544-0
35. Saco PM, Rodríguez JF, Moreno-de las Heras M, Keesstra S, Azadi S, Sandi S, Baartman J, Rodrigo-Comino J, Rossi MJ (2020) Using hydrological connectivity to detect transitions and degradation thresholds: applications to dryland systems. CATENA 186(February):104354. https://doi.org/10.1016/j.catena.2019.104354
36. Snelder TH, Cattanéo F, Suren AM, Biggs BJF (2004) Is the River Environment Classification an improved landscape-scale classification of rivers? J N Am Benthol Soc 23(3):580–598. https://doi.org/10.1899/0887-3593(2004)023%3c0580:ITRECA%3e2.0.CO;2
37. Turnbull L, Wainwright J (2019) From structure to function: Understanding shrub encroachment in drylands using hydrological and sediment connectivity. Ecol Indicat 98(March 2018):608–618. https://doi.org/10.1016/j.ecolind.2018.11.039
38. USGS (2020) Sediment diagram, from source to sink. https://www.usgs.gov/media/images/sediment-diagram-source-sink
39. Wohl E (2017) Connectivity in rivers. Progr Phys Geogr Earth Environ 41(3):345–362. https://doi.org/10.1177/0309133317714972
40. Wohl E (2020) Rivers in the landscape. Wiley. https://doi.org/10.1002/9781119535409
41. Wohl E, Brierley G, Cadol D, Coulthard TJ, Covino T, Fryirs KA, Grant G, Hilton RG, Lane SN, Magilligan FJ, Meitzen KM, Passalacqua P, Poeppl RE, Rathburn SL, Sklar LS (2019) Connectivity as an emergent property of geomorphic systems. Earth Surf Proc Land 44(1):4–26. https://doi.org/10.1002/esp.4434
42. Wohl E, Magilligan FJ, Rathburn SL (2017) Introduction to the special issue: connectivity in geomorphology. Geomorphology 277:1–5. https://doi.org/10.1016/j.geomorph.2016.11.005
43. Yang X, Lu X, Ran L, Tarolli P (2019) Geomorphometric assessment of the impacts of dam construction on river disconnectivity and flow regulation in the Yangtze Basin. Sustainability 11(12):3427. https://doi.org/10.3390/su11123427
44. Yu Y, Feng J, Liu H, Wu C, Zhang J, Wang Z, Liu C, Zhao J, Rodrigo-Comino J (2023) Linking hydrological connectivity to sustainable watershed management in the Loess Plateau of China. Current Opin Environ Sci Health 35(July):100493. https://doi.org/10.1016/j.coesh.2023.100493
45. Zhang Y, Huang C, Zhang W, Chen J, Wang L (2021) The concept, approach, and future research of hydrological connectivity and its assessment at multiscales. Environ Sci Pollut Res 28(38):52724–52743. https://doi.org/10.1007/s11356-021-16148-8
46. Zhao G, Gao P, Tian P, Sun W, Hu J, Mu X (2020) Assessing sediment connectivity and soil erosion by water in a representative catchment on the Loess Plateau, China. CATENA 185(September):104284. https://doi.org/10.1016/j.catena.2019.104284

Chapter 7
Land Use Land Cover Dynamics and Sediment Variation

Abstract The Damodar River Basin (DRB), encompassing the Maithon, Panchet, and Tenughat Dam catchments, has experienced substantial Land Use Land Cover (LULC) changes that influence soil erosion and sediment dynamics. This chapter explores the relationship between LULC variations and sediment yield, with a focus on anthropogenic impacts such as urbanization, agriculture, and deforestation. These land cover changes contribute significantly to soil erosion and sediment transport in the catchments. The research compares soil erosion, sediment yield, and sediment connectivity across the three dams, providing a comprehensive understanding of their varying impacts. Additionally, the chapter evaluates the water-holding capacity and trapping efficiency of each dam, highlighting their effectiveness in controlling sediment inflow. The chapter findings underscore the importance of monitoring LULC dynamics to manage soil erosion and sedimentation in dam catchments, offering key insights for sustainable watershed management. This research aims to improve strategies for sediment control and reservoir management in the region.

Keywords Land use dynamics · Anthropogenic impact · Trapping efficiency · Water holding capacity · Hypothesis · Soil erosion · Sediment yield

7.1 Introduction

Land use and land cover (LULC) maps serve as a dynamic resource for analyzing the interactions between human activities and the natural environment [23, 40]. The LULC dynamics in any region are the reflection of a positive correlation between the demand for land and the overgrowing population pressure [2]. Therefore, human activity and land use dynamics have intense implications on soil resources [3, 33] which intensify the soil erosion and sediment yield in the catchment or watershed. Various researchers across the globe focus on the topic of land use and land cover impacts on land resources [1, 24, 25, 33]. The Damodar River basin stands as a watershed profoundly influenced by anthropogenic activities, which markedly exacerbate soil erosion and amplify sediment yield. Intensive mining activities such as open-cast

S. A. Alam and R. Maiti, *Sediment Source to Sink: Deciphering Sediment Connectivity to Large Dams in Damodar River Basin*, Earth and Environmental Sciences Library,
https://doi.org/10.1007/978-3-031-87072-9_7

and underground mining along with industrial operations, significantly alter land use and land cover dynamics leading to environmental degradation [34]. Thus, sediment production from the upper catchment has steadily increased over time, driven by the escalating rate of anthropogenic impacts. Recent studies have also shown that combining land use changes and extreme climate events can be imperative in understanding sediment yield at key sites, particularly concerning sediment deposition and erosion [6].

Land use land cover change can lead to significant soil erosion and sediment yield in the watershed. The accumulating sediments significantly impact reservoir storage capacity, ultimately reducing the reservoir's operational lifespan. Researchers examined the impact of erosion-driven sediment yield on reservoir storage capacity [19], karst region reservoir storage [39], and reservoir sedimentation [18, 28, 37]. The storage capacities of the major dams in the Damodar River basin (Maithon, Panchet, and Tenughat) have been consistently declining due to the significant sediment influx from the upper catchment. The decline in reservoir storage capacity is predominantly site-specific and governed by the regional characteristics of the basin [37]. Therefore, a thorough understanding of regional characteristics is essential for accurately evaluating the factors influencing sediment production intensity in a specific area. Soil erosion and spatial sediment yield are continuously increasing due to the combined effects of intensification of rain and land use changes driven by human activities [6, 29]. To address the issues mentioned above, we combined LULC dynamics and spatial sediment yield to investigate variations in reservoir sedimentation and the contribution of land use changes to sediment production. Additionally, we formulated a hypothesis to assess which reservoir is most affected by sedimentation and its connectivity to the sediment production zone. The results presented in this chapter will enhance our understanding of how sediment production from LULC dynamics and sediment connectivity influence the rate of sedimentation in the reservoir.

7.2 Land Use Land Cover (LULC) Dynamics in Dam Catchments

The Land Use Land Cover (LULC) dynamics within the Maithon, Panchet, and Tenughat dam catchments reveal significant transformations, highlighting the influence of human activities and environmental shifts. A substantial decline in agricultural land and forest cover indicates the impacts of deforestation, land degradation, and shifting land-use priorities. Concurrently, the rapid expansion of built-up areas highlights intensified urbanization and infrastructure development, while the increase in fallow land and reduction in forest areas point to altered land management practices and potential sedimentation challenges. These findings emphasize the critical need for sustainable land-use strategies to address ecological degradation, sediment management, and the long-term resilience of these dam catchments (Maithon, Panchet, and Tenughat).

7.2.1 Maithon Dam Catchment

The land use and land cover (LULC) dynamics in the Maithon Dam catchment area between 1995 (Fig. 7.1) and 2024 (Fig. 7.2) exhibit profound transformations indicative of significant anthropogenic and sediment variations (Table 7.1). Agricultural land has contracted sharply from 3890 to 2729 km^2, reflecting changes in land utilization driven by urbanization or land abandonment. Built-up areas have expanded exponentially, from a mere 76–789 km^2, underscoring the region's rapid urban growth and infrastructural development. Fallow land has surged from 748 to 2069 km^2, suggesting a rapid transition in cultivation patterns or land degradation over time. The decline in forested areas, with tree cover reducing from 1537 to 733 km^2, highlights deforestation and the associated loss of ecological resilience. Additionally, water bodies have diminished from 216 to 142 km^2, potentially due to sedimentation, landfilling, water extraction, or climate variability. These findings underscore the urgency for sustainable land-use planning and ecosystem restoration initiatives to mitigate the environmental and hydrological impacts of LULC changes on the Maithon Dam catchment and its surrounding areas.

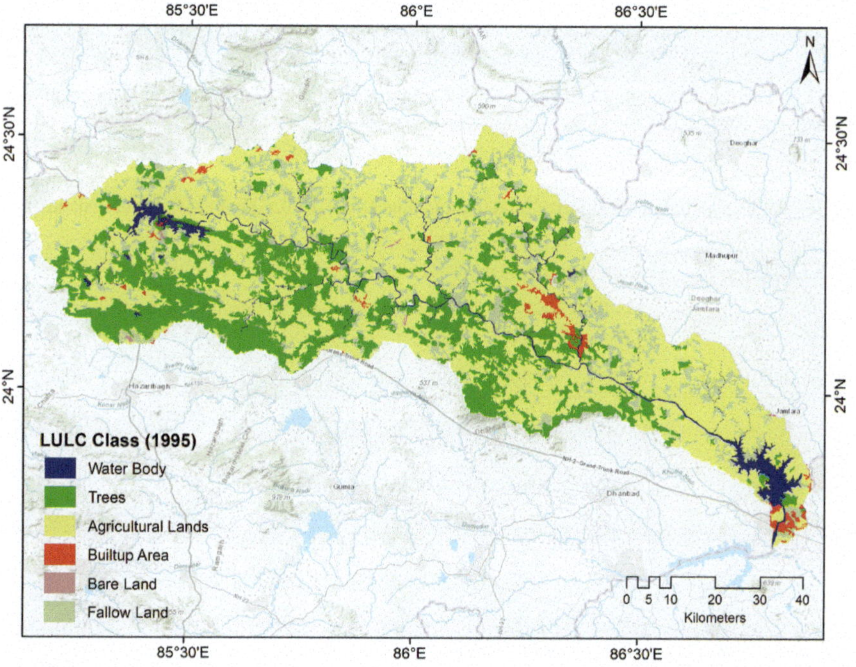

Fig. 7.1 The Land Use and Land Cover (LULC) distribution map of the Maithon Dam catchment for the year 1995

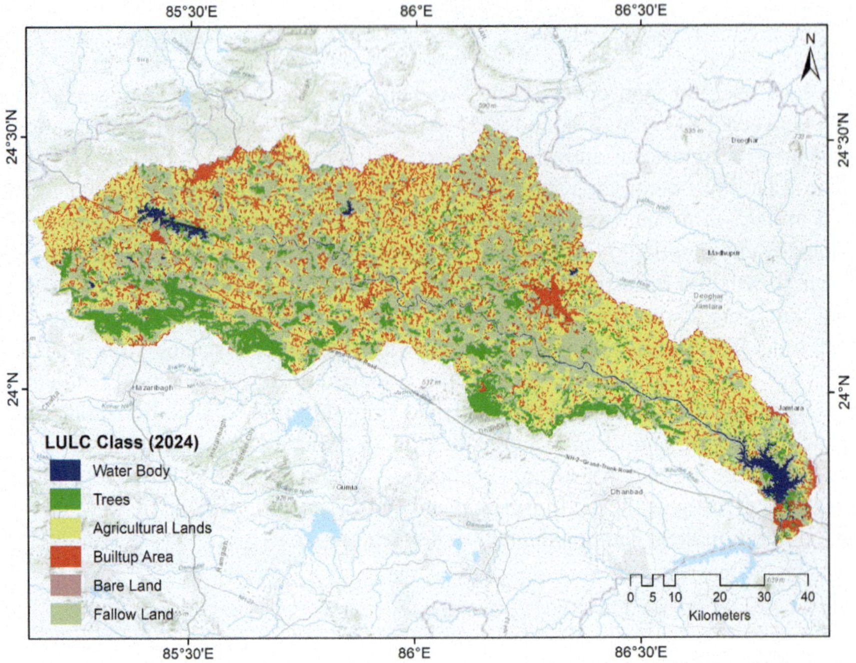

Fig. 7.2 The Land Use and Land Cover (LULC) distribution map of the Maithon Dam catchment for the year 2024

Table 7.1 The spatial distribution of land use and land cover (LULC) in the Maithon, Panchet, and Tenughat dam catchments for the years 1995 and 2024

LULC class	Area (km^2)					
	Maithon Dam		Panchet Dam		Tenughat Dam	
	1995	2024	1995	2024	1995	2024
Agricultural lands	3890	2729	4203.08	2805.36	1715.37	1340.12
Bare land	7	12	5.65	10	25.65	40.89
Built-up area	76	789	335.30	891.69	126.24	350.64
Fallow land	748	2069	408.25	1526.61	568.35	1485.75
Trees	1537	733	1565.89	1323.97	2120.38	1353.97
Water body	216	142	185.43	145.99	92.30	76.20

7.2.2 Panchet Dam Catchment

The Land Use and Land Cover (LULC) dynamics in the Panchet Dam catchment between 1995 (Fig. 7.3) and 2024 (Fig. 7.4) reveal significant shifts in land use patterns (Table 7.1), indicative of both anthropogenic activities and ecological variations. Agricultural land reduced substantially from 4203.08 km^2 in 1995 to 2805.36

km^2 in 2024, suggesting a notable transition in land use, potentially driven by urban expansion, land degradation, rehabilitation, abandonment due to land subsidence, or changes in farming practices. Built-up areas have expanded considerably, increasing from 335.30 to 891.69 km^2, reflecting the intensification of urbanization and infrastructure development. Conversely, fallow land has increased dramatically from 408.25 to 1526.61 km^2, which may be attributed to agricultural abandonment or shifts in cultivation practices. The area covered by trees has declined from 1565.89 to 1323.97 km^2, highlighting a reduction in forest cover and potential environmental degradation. Additionally, water bodies have slightly decreased from 185.43 to 145.99 km^2, likely due to sedimentation, water extraction, or other hydrological changes. These findings highlight the urgency for sustainable land management practices to mitigate the effects of urbanization, deforestation, and land abandonment, ensuring the long-term ecological and hydrological health of the Panchet Dam catchment.

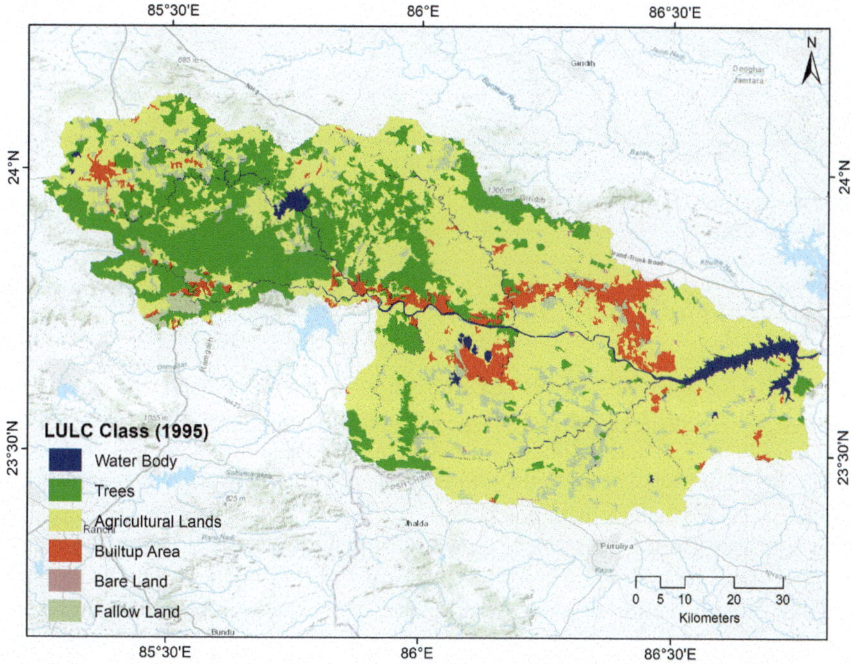

Fig. 7.3 The Land Use and Land Cover (LULC) distribution map of the Panchet Dam catchment for the year 1995

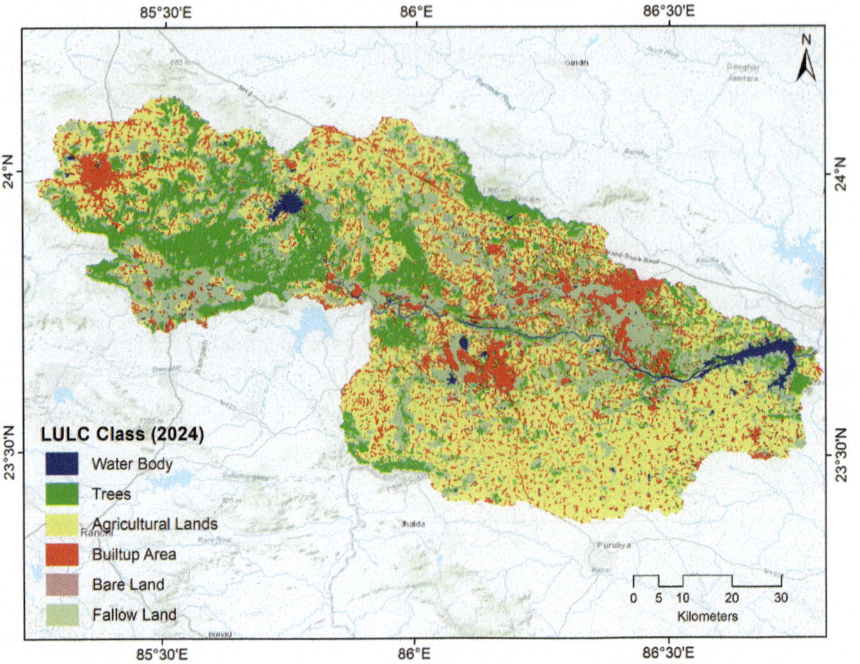

Fig. 7.4 The Land Use and Land Cover (LULC) distribution map of the Panchet Dam catchment for the year 2024

7.2.3 *Tenughat Dam Catchment*

The Land Use and Land Cover (LULC) shifts in the Tenughat Dam catchment between 1995 (Fig. 7.5) and 2024 (Fig. 7.6) indicate significant alterations in both land use (Table 7.1) and sediment yield dynamics. Agricultural land has decreased from 1715.37 to 1340.12 km^2, reflecting a decline in active cultivation, potentially due to changes in land management practices, agricultural abandonment, or a transition to other land uses. Built-up areas have grown substantially, from 126.24 to 350.64 km^2, pointing to accelerated urbanization and infrastructure development, which may have led to the fragmentation of natural landscapes. The expansion of fallow land from 568.35 to 1485.75 km^2 suggests a shift in agricultural activity, possibly due to reduced agricultural productivity or changes in land tenure systems. Furthermore, the reduction in tree cover from 2120.38 to 1353.97 km^2 signals a notable loss of forested areas, which could have implications for biodiversity conservation, water retention, and sediment control within the catchment. The slight decrease in water bodies from 92.30 to 76.20 km^2 reflects potential hydrological alterations, including sedimentation and reduced water storage capacity. These findings emphasize the importance of adopting integrated land management strategies to address the challenges posed

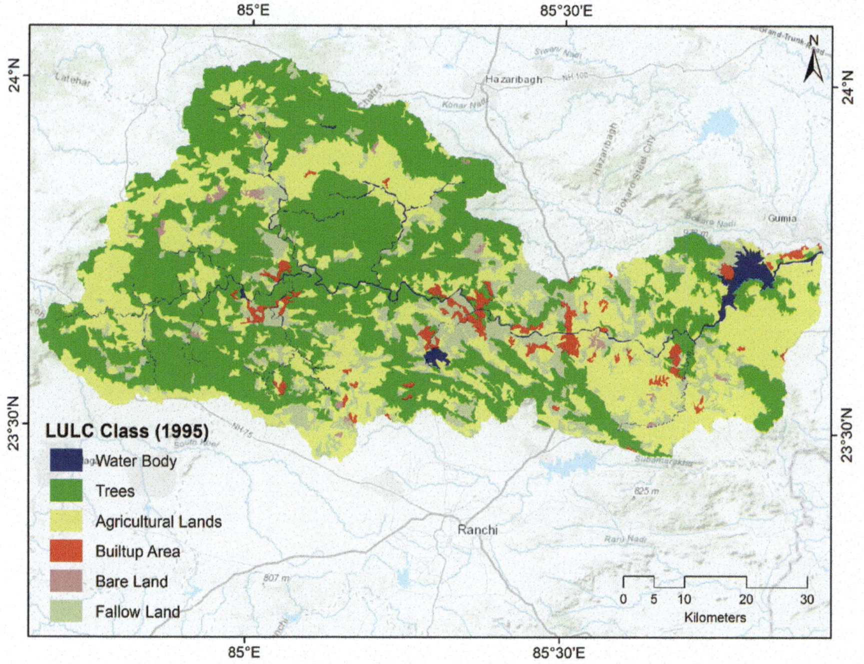

Fig. 7.5 The Land Use and Land Cover (LULC) distribution map of the Tenughat Dam catchment for the year 1995

by urbanization, land degradation, and ecological loss, ensuring the sustainability of the Tenughat Dam catchment's hydrological and environmental functions.

7.3 Accuracy Assessment

The accuracy assessment of a land use/land cover map depends on the quantity of input data, the quality of field observations, and the resolution of the satellite sensor. It is typically evaluated using metrics such as overall accuracy and the kappa coefficient [30, 38, 40]. In our investigation, we utilized land use/land cover (LULC) data products from NASA Earthdata for 1995 and ESRI datasets for 2024 [14, 30]. The NASA Earthdata LULC product for 1995 exhibited strong performance, with an overall accuracy of 94.46% and a kappa coefficient of 0.9445, indicating high reliability in classifying land cover types for that period [30]. In contrast, a 2024 study examining three recent 10-m global land cover maps found lower overall accuracy, ranging from 73.4 to 83.8% [14]. Despite the differences in accuracy, both datasets are valuable for understanding land cover dynamics, which are crucial for assessing sediment production in various regions. The high accuracy of the 1995 NASA Earthdata LULC product enables us to confidently analyze long-term land

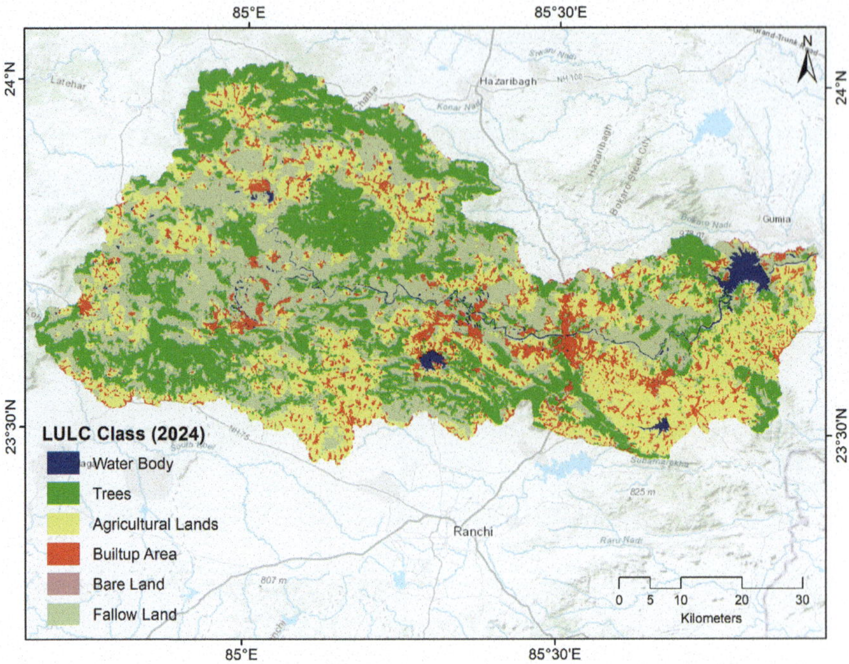

Fig. 7.6 The Land Use and Land Cover (LULC) distribution map of the Tenughat Dam catchment for the year 2024

cover changes, while the more recent ESRI datasets offer insight into current land use trends. Together, these datasets provide a comprehensive view of land cover transitions over time, supporting the analysis of environmental changes such as sediment yield and land degradation.

7.4 Importance of Studying LULC Dynamics and Its Relationship with Soil Erosion and Sediment Yield in the Dam Catchments

Understanding the dynamics of land use and land cover (LULC) in the Maithon, Panchet, and Tenughat dam catchments between 1995 and 2024 (Fig. 7.7) is crucial for assessing environmental sustainability and sediment management in these regions. Across all three catchments, a consistent trend of declining agricultural lands is observed, with substantial portions transitioning to built-up areas and fallow lands. For instance, in the Maithon catchment, 2160 km^2 of agricultural land was retained, while 630 and 957 km^2 were converted into built-up and fallow lands, respectively. Similarly, in Panchet, 2434 km^2 of agricultural land remained, but significant areas transitioned to fallow land (825 km^2) and urbanized regions (615 km^2). The Tenughat catchment shows a parallel pattern, with 962 km^2 of agricultural land preserved, yet 450 km^2 and 207 km^2 shifted to fallow and built-up areas. Forested areas have also undergone considerable transformation; Maithon and Panchet experienced notable transitions from tree cover to fallow land and urban areas, while Tenughat saw a reduction of forested areas with considerable shifts to agricultural and fallow categories. Water bodies in all catchments showed considerable reduction, primarily transitioning to adjacent categories such as agricultural and fallow lands, indicating potential sedimentation or hydrological alterations.

The LULC transformations contributing most significantly to soil erosion and sediment yield in the Maithon, Panchet, and Tenughat dam catchments are the conversions of agricultural lands and forested areas to fallow lands and built-up areas. These transformations exacerbate sediment yield, as the loss of dense vegetation accelerates the detachment and movement of soil particles during rainfall events. Changes such as the decline in agricultural lands and forest cover, coupled with the expansion of fallow and built-up areas, significantly influence hydrological processes and sedimentation rates. These dynamics affect water resource availability, reduce reservoir storage capacity, and impair flood control and hydropower generation. By analyzing LULC changes, stakeholders can develop targeted watershed management strategies, including afforestation and sustainable urban planning, to mitigate erosion, manage sediment transport, and ensure the long-term resilience of these critical water infrastructures.

The Land Use and Land Cover (LULC) dynamics in the Maithon, Panchet, and Tenughat dam catchments reveal significant trends in soil erosion and sediment yield across different classes from 1995 to 2024 (Fig. 7.8). In all three catchments, bare land and fallow land consistently exhibit the highest soil erosion and sediment yield values, highlighting their vulnerability to surface runoff and sediment transport. Conversely, water bodies display the lowest values, emphasizing their limited role in sediment generation but significant impact on sediment deposition and flow regulation. In the Maithon catchment, a notable increase in soil erosion from agricultural lands (2.87–3.15 t/ha/year) and fallow land (7.47–9.59 t/ha/year) is the contribution of intensified land-use changes and possible land degradation due to reduced vegetation cover. Similarly, the Panchet catchment shows an escalation in erosion rates in bare land (10.55–12.36 t/ha/year) and fallow land (9.07–11.56 t/ha/year), suggesting inadequate land management practices and the expansion of degraded zones. The Tenughat catchment demonstrates a similar pattern, with bare land erosion peaking at 14.56 t/ha/year in 2024, underscoring its critical role in sediment yield contributions to downstream reservoirs. These findings align with ground realities in the dam catchments, where anthropogenic activities such as deforestation, agricultural expansion, and unplanned urbanization have exacerbated soil erosion. Increased sediment loads in these reservoirs threaten storage capacities, water quality, and ecological stability. Therefore, various management practices in these catchments are imperative to reduce soil erosion and maintain the long-term viability of the dams in the Damodar River Basin.

Fig. 7.7 Chord diagrams depicting land use and land cover (LULC) transitions from 1995 to 2024 for (a) Maithon Dam catchment, (b) Panchet Dam catchment, and (c) Tenughat Dam catchment

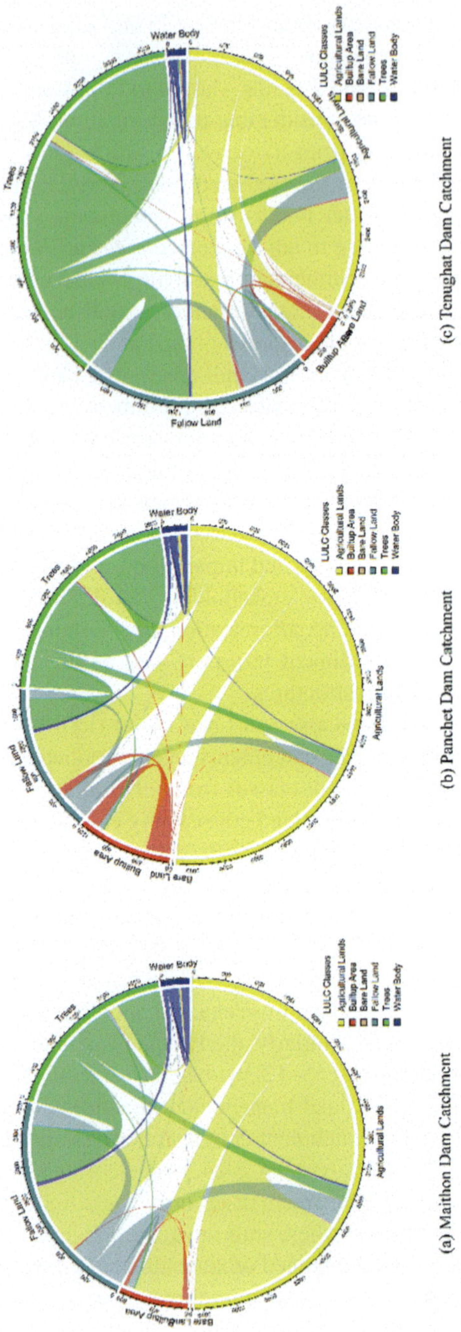

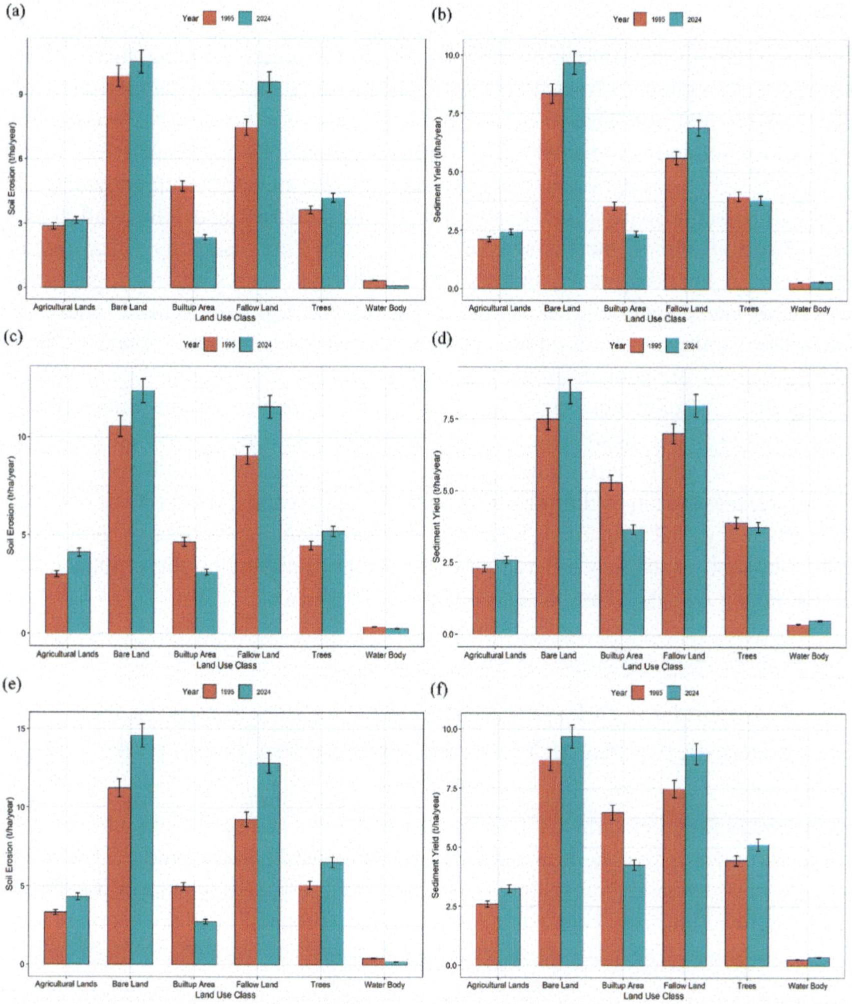

Fig. 7.8 Soil erosion across various land use classes in the years 1995 and 2024 is represented for the (a) Maithon Dam catchment, (c) Panchet Dam catchment, and (e) Tenughat Dam catchment, while sediment yield for the same land use classes and years is depicted for the (b) Maithon Dam catchment, (d) Panchet Dam catchment, and (f) Tenughat Dam catchment

7.5 Anthropogenic Impact on Sediment Yield and Sediment Connectivity

The Maithon, Panchet, and Tenughat dam catchments in the Chota Nagpur Plateau have experienced significant anthropogenic impacts, altering sediment production and sediment connectivity. Large-scale open-cast and underground mining operations, involving activities such as drilling, blasting, coal washing, and excavation, are primary causes of soil erosion and sediment generation in these regions [21]. The Damodar River's tributaries run through mining areas including Jharia, North Kanakpur, Raniganj, Bokaro, Potanga, Bhurkunda, Ghatotand, Pankri Barwadih, Giridh, Koderma, and Hazaribagh coalfields. This is why the river has become one of the most polluted in India [31]. Catchments dominated by mining activities in this region are highly susceptible to severe soil erosion and elevated sediment yield [27]. Rapid urbanization, agricultural expansion, mining activities, and deforestation have intensified soil erosion, increasing sediment inflow to these reservoirs. Infrastructure development, including roads and settlements, disrupts natural sediment pathways, causing localized sediment deposition or detachment. Sediment transfer within the catchment exhibits spatio-temporal variability, marked by irregular patterns along pathways from the source areas to the outlet [5, 15, 16, 26]. Improper land-use management and unplanned construction intensify sediment disconnection, reducing sediment delivery to downstream systems. Deforestation in upstream areas accelerates erosion, while agricultural practices in slope regions disturb soil stability. Therefore, these changes diminish reservoir storage capacity, reduce sediment trapping efficiency, and jeopardize water resource sustainability. Effective monitoring and sustainable management strategies are crucial to mitigating these impacts and strengthening watershed and reservoir capacity resilience.

7.6 Comparison Between Major Dams Considering Soil Erosion, Sediment Yield, and Sediment Connectivity

The analysis of soil erosion (SE), sediment yield (SY), and sediment pathways (IC) for all the major dams (Maithon, Panchet, and Tenughat) in the Damodar River Basin helps us to understand their sediment dynamics, reservoir sedimentation challenges, and the critical interplay between upstream erosion processes and downstream sediment deposition. The analysis of sediment dynamics from 1990 to 2020 reveals notable shifts in the sediment Connectivity (IC), Soil Erosion (SE), and Sediment Yield (SY), reflecting the evolving geomorphological and hydrological processes in the study region. A violin plot depicts the distribution and variability of data points' frequency within each region (Fig. 7.9). The negative values of sediment connectivity indicate disconnected areas and positive values may indicate well-connected areas. The IC values exhibit both negative and positive trends over time, reflecting the spatial as well as temporal variability of sediment connectivity influenced by diverse anthropogenic impacts and natural factors such as elevation, slope, soil, and vegetation for all the dams. The SE and SY values demonstrate both increasing and decreasing trends over time and space, highlighting the Spatiotemporal heterogeneity of erosion and sediment production across the entire catchment. This variability is attributed to changes in LULC, management practices, anthropogenic impacts such as mining activities, and natural factors, which collectively influence erosion, sediment production, and sediment pathways throughout the catchment.

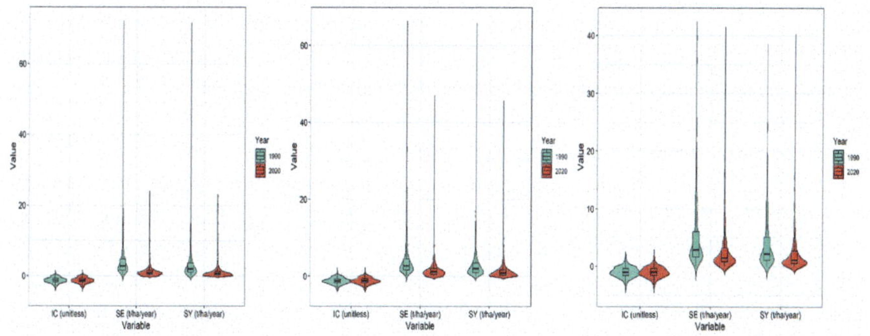

(a) Maithon Dam Catchment (b) Panchet Dam Catchment (c) Tenughat Dam Catchment

Fig. 7.9 Violin plots of sediment connectivity (IC), soil erosion (SE), and sediment yield (SY) for the major dams (Maithon, Panchet, and Tenughat) in the Damodar River basin during the period from 1990 to 2020

The Maithon dam catchment and Panchet dam catchment, characterized by moderate connectivity, experienced higher sediment production in 1990 due to enhanced erosion processes. However, by 2020, peak soil erosion and peak sediment production had reduced but total areal coverage expanded (Fig. 7.9a). Significant increases in SY in some regions may point to localized factors like intensified land use changes or insufficient sediment control measures. In the Tenughat dam catchment, consistent erosion and improved connectivity from 1990 to 2020 indicate a steady supply of sediment delivery to the outlets. The areal expansion of erosion and connectivity is evident across all dams in 2020 (Fig. 7.9). Consequently, the mixed trend in sediment production reflects localized variations driven by land use changes, hydrological shifts, and sediment retention strategies. These findings underscore the interplay between connectivity, erosion, and sediment delivery processes, emphasizing the critical role of targeted watershed management and erosion control measures to address sedimentation challenges while ensuring sustainable landscape management.

7.7 Water Holding Capacity of the Major Dams

Reservoirs are being constructed worldwide at an increasing rate to control flooding in lower catchments, supply water, support irrigation, and generate hydropower [35]. The water holding capacity of a reservoir is important to maintain its significant role in regional water management, flood control, and power generation. Reservoir sedimentation is negatively correlated with water-holding capacity. The sediment produced in the upper catchment due to intensive soil erosion is a major contributor to sedimentation in reservoirs. The more pronounced the sediment yield in the upper catchments, the greater the reduction in storage capacity over time. Therefore, onsite issues, such as agricultural problems and land degradation, contribute to this process, while off-site effects primarily involve sedimentation in reservoirs and the corresponding loss of water storage capacity [20, 36]. The cumulative loss in storage capacity of the dams is closely linked to soil erosion and sediment yield in the upper catchments. The Maithon Dam, with a gross storage capacity of 1093.54 million m^3 and an effective storage of 77.53 million m^3 (Table 7.2), serves similar purposes but has experienced a higher storage loss of 27.4% by 2001 [7, 9] due to its longer operational period. The Panchet Hill Dam, with an even greater gross storage capacity of 1193.46 million m^3 and an effective storage of 748.38 million m^3 (Table 7.2), plays a critical role in flood management and water supply. Despite its larger size, it has seen a 25.17% storage loss by 1995 [7, 9], which could impact its long-term operational efficiency. Finally, the Tenughat Dam, with the highest effective storage of 999.92 million m^3 from a gross storage of 1,020.9 million m^3 (Table 7.2), plays an essential role in water supply and power generation. The cumulative loss in storage capacity is relatively less (16.8% by 2001) [7, 9], reflecting better maintenance or less sedimentation. Although storage loss due to sedimentation, becomes more critical in regional water resource management, flood control, irrigation, and hydropower production in a situation of climate change when intensified rain becomes more frequent.

7.8 Trapping Efficiency

Reservoir sedimentation is a significant problem worldwide, particularly in tropical countries where sediment production rates are exceptionally high. This problem directly affects the performance of the reservoir and its life span. The capacity of a reservoir is inversely related to the amount of sediment deposited. Therefore, the Trapping efficiency (TE) method is important for understanding the sedimentation dynamics of reservoirs and their impact on water resource management. There are several methods of estimated trapping efficiency (TE). The TE method was proposed by [4] and was developed by [17] which is the percentage of sediment retained by a reservoir is determined using an empirical relationship based on the capacity-to-inflow (C/I) ratio represented by standard efficiency curves. The method starts from the original equation [4]:

$$Te = 1 - \frac{0.5}{\sqrt{\Delta\tau_\Gamma}} \tag{7.1}$$

where:

Te Trap efficiency (% in decimal units),

$\Delta\tau_\Gamma$ The local residence time changes in a year

Table 7.2 A detailed description of the storage capacity and cumulative storage loss of five major dams in the Damodar River basin [7, 9]

Name of dam	KONAR	MAITHON	PANCHET	TILAIYA	TENUGHAT
Operated and maintained by	DVC	DVC	DVC	DVC	WRD, Jharkhand
Longitude of dam	85°46′ E	86°49′ E	86°46′ E	85°31′ E	85°48′ E
Latitude of dam	23°56′ N	23°47′ N	23°40′ N	24°19′ N	23°45′ N
Year of completion	1955	1957	1959	1953	1978
Basin	Damodar	Damodar	Damodar	Damodar	Damodar
River	Konar	Baraka	Damodar	Barakar	Damodar
Nearest city	Hazaribagh	Asansol	Asansol	Kodarma	Hazaribagh
Seismic zone	3	3	3	3	3
Type	Earth, Masonry	Earth, Masonry	Earth, Masonry	Masonry	Earth
Height above lowest foundation	57.60 m	Concrete dam 56.08 m Earth dam 49.99 m	Concrete dam 47.85 m Earth dam 40.84 m	30.18 above river bed level	64.87 m
Length of dam	Concrete + Earth dam 3682.03 m (277.37 + 3404.66 m)	Concrete + Earth dam 4426.76 m (362.41 + 4064.35 m)	Concrete + Earth dam 6777 m (370 + 6407 m)	365.76 m	6494.00
Volume content of dam (1000 m^2)	Concrete volume 322. 602 and Earth volume 4166.847	Concrete volume 243. 698 and Earth volume 3475.685	Concrete volume 258. 00 and Earth volume 5267.00 m	108.76	
Gross storage capacity (1000 m^3)	247,960 (Upto EL 427.94 m)	1,093,540 (up to EL 152.40 m)	1,193,464 (up to EL 135.64 M	380,710 (up to EL 372.47 m)	1,020,900.00
Reservoir area (1000 m^2)	25,900 (upto EL427.94 m)	107,160 (up to EL 152.40 m)	121,386 (up to EL 135.64 m)	59,080 (up to EL 372.47 m)	64,780.00
Effective storage capacity (1000 m^3)	213,560.00	77,527 (up to EL 150.88 M)	748,375 (up to EL 132.59 m)	305,930.00	999,920.00

(continued)

Table 7.2 (continued)

Name of dam	KONAR	MAITHON	PANCHET	TILAIYA	TENUGHAT
Purpose	Irrigation, Hydel Power, Flood Control, Water supply	Irrigation, Hydel Power, Flood Control, Water Supply	Irrigation, Hydel Power, Flood Control, Water Supply	Irrigation, Hydel Power, Flood Control, Water Supply	Irrigation, Hydel Power, Water Supply
Designed spillway capacity (m^3/s)	6796.00	13,592.00	16,608.00	1348.00	16,000.00
Cumulative loss of storage capacity (%)	25.43% till 1996 (CWC)	27.40% till 2001 (DVC)	25.17% till 1995 (DVC)	36.01% till 1997 (CWC)	16.8% till 2001 (CWC)

$$\Delta\tau_{\Gamma} = \frac{\sum_{i=1}^{nj} Vi}{I} \tag{7.2}$$

where,

V Storage capacity of the reservoir in km^3
I Inflow volume in km^3/day.

The TE was developed and called one of the most important properties of a reservoir. The developed method is generally defined as the ratio of the total sediment retained by a reservoir to the total sediment that enters the reservoir [17] as expressed by the following equation:

$$Te = \frac{V_i - V_o}{V_i} \times 100 \tag{7.3}$$

where,

Te Trap efficiency,
V_i The Inflow sediment load,
V_o The Outflow sediment load,

The method outlined in Eqs. 7.1 and 7.3 uses the reservoir capacity-to-inflow ratio to determine the trapping efficiency for major reservoirs in the Damodar River basin. The analysis of the trapping efficiency (TE) for the five major reservoirs in the Damodar River Basin—Konar, Maithon, Panchet, Tilaiya, and Tenughat—reveals significant variations in sediment retention capacities. Konar and Panchet Reservoirs achieve the highest TE of 93% (Table 7.3), benefiting from their upstream positions, moderate inflow volumes, and steep catchment slopes that enhance sediment deposition within the reservoirs. Maithon, with a TE of 92% (Table 7.3), balances high

Table 7.3 Trapping efficiency (TE) of the five major reservoirs in the Damodar River Basin (data compiled from various sources [8, 10, 11–13])

Reservoir	Full reservoir level (m)	Average inflow (MCM)	Capacity inflow ratio	Trapping efficiency (%)
Konar	429.20	506.09	0.493	93
Maithon	150.9	2926.31	0.215	92
Panchet	135.6	3808.692	0.335	93
Tilaiya	372.46	3076.21	0.123	88
Tenughat	268.834	1750.9	0.308	90

inflow volumes from tributary confluences with structural designs suited for multi-purpose functions. In contrast, Tilaiya, with the lowest TE of 88%, struggles with higher sediment loads from its rugged terrain catchment and a low capacity-to-inflow ratio, limiting its retention ability. Tenughat, with a TE of 90%, reflects moderate efficiency due to its midstream position, where flatter terrain facilitates sedimentation but upstream contributions remain significant. These variations highlight the interplay of geomorphology, inflow dynamics, and reservoir design, emphasizing the need for tailored sediment management approaches to sustain operational efficiency across the basin. These findings emphasize the need for sediment management and monitoring to sustain reservoir capacity and functionality.

7.9 Formulating the Hypothesis

The hypothesis is a tentative solution to a problem [32]. A hypothesis is a statement that addresses a specific research question and predicts the expected outcomes of an experiment which means a mere assumption or some supposition to be proved or disproved [22]. In this investigation, we formulate a research hypothesis to explore the optimal correlation between sediment connectivity pathways and sediment yield across the catchments of the Maithon, Panchet, and Tenughat dams in the Damodar River basin and their respective reservoirs. The hypothesis can be broadly classified into two categories: (a) Null Hypothesis (H_0) and (b) Alternative Hypothesis (H_1). Therefore, our research hypothesis is as follows:

H_0 = There is no significant difference in the correlation between sediment connectivity and sediment yield across the catchments of the Maithon, Panchet, and Tenughat dams.

H_1 = There is a significant difference in the correlation between sediment connectivity and sediment yield across the catchments of the Maithon, Panchet, and Tenughat dams.

This hypothesis is tested through ANOVA (Analysis of Variance) to evaluate how well sediment connectivity relates to sediment yield across these three dams, providing insights into sediment management strategies for the region. Based on this test, we can determine the degree of connectivity and distinguish their unique characteristics.

7.10 ANOVA Test

Analysis of variance (ANOVA) is a statistical test used to assess the difference between the means of more than two groups. The F statistic compares the variability between the groups to the variability within the groups. The ANOVA method is expressed by the following equation:

$$F = \frac{MST}{MSE} \tag{7.4}$$

$$MST = \frac{\sum_{i-1}^{k}\left(T_i^2/n_i\right) - G^2/n}{k-1} \tag{7.5}$$

$$MSE = \frac{\sum_{i-1}^{k}\sum_{j-1}^{n_i} Y_{ij}^2 - \sum_{i-1}^{k}\left(T_i^2/n_i\right)}{n-k} \tag{7.6}$$

where,

- F Variance ratio for the overall test,
- MST Mean square due to treatments (Between groups),
- MSE Mean square due to error (within groups),
- Y_{ij} Observations,
- T_i Group total,
- G Total for all observations,
- n_i number in group i,
- n Total number of observations.

The ANOVA test results represent the statistically significant (F = 8.55, P = 0.000388) differences (Fig. 7.10) within the group (Table 7.4). The post-hoc test shows that both the Bonferroni and Tukey tests indicate no statistically significant difference between the two dams, with p-values of 0.092461 and 0.0777, respectively. The null hypothesis (H_o) of no difference is failed to be rejected (FTR H_o). Similarly, the p-values for both Bonferroni (0.165678) and Tukey (0.13288) tests suggest no significant difference between Maithon and Tenughat. The null hypothesis remains valid (FTR. H_o). Both tests revealed a highly significant difference between Panchet and Tenughat, with Bonferroni and Tukey p-values of 2.34E−0.04 and 2.28E−04, respectively. The null hypothesis is rejected (Reject H_o), confirming that the two dams differ significantly in their characteristics. The analysis shows that Maithon and Panchet, as well as Maithon and Tenughat, exhibit no statistically significant differences ($p > 0.05$) in their tested variables. The post-hoc test represents

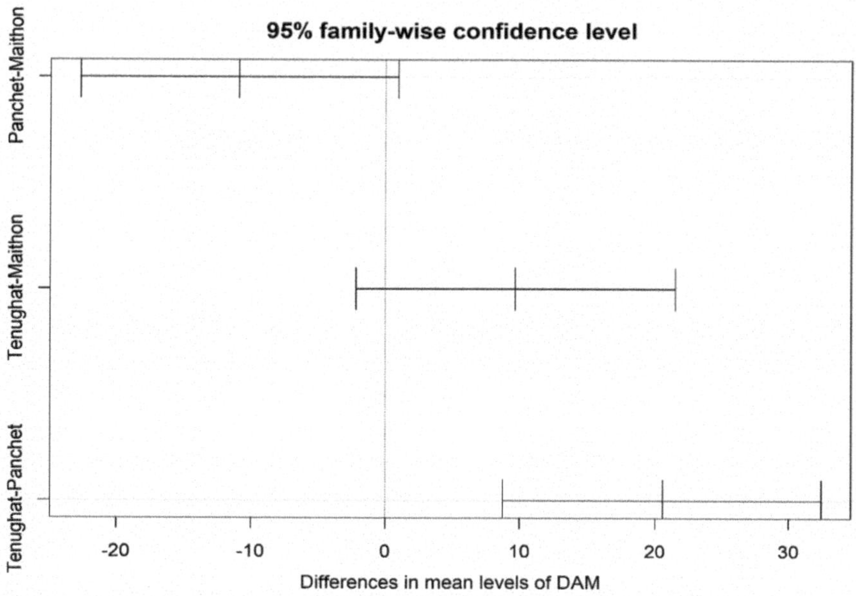

Fig. 7.10 The figure represents the mean difference within the group (Tukey test)

Table 7.4 The result of the one-way ANOVA

	Df	Sum Sq	Mean Sq	F value	Pr (> F)
Dam	2	6794	3397	8.556	0.000388***
Residuals	93	36,921	397		

Signif. codes: *** 0.001

the statistically significant differences ($p < 0.001$) between the Panchet and Tenughat Dams (Table 7.5). Both Maithon and Panchet dams are situated within the Damodar River Basin, sharing similar geomorphological and climatological characteristics. The discharge and sediment dynamics in the Panchet Dam are significantly influenced by the Tenughat Dam and Konar Dam. The high intensity of mining activities in the Tenughat Dam catchment substantially contributes to sediment production, resulting in a significant difference ($p < 0.001$) between the Tenughat and Panchet dams. This difference highlights that the Tenughat Dam is well-connected to its upper catchment emphasizing its critical role in sediment management and catchment dynamics within the basin.

7.11 Major Findings

The findings in this chapter are as follows.

Table 7.5 Post-hoc test result for the three major dams

Group	"Bonf" test	"Tukey" test		
	P-value	Decision	*P*-value	Decision
Maithon versus Panchet	0.092461	FTR H_o	0.0777	FTR H_o
Maithon versus Tenughat	0.165678	FTR H_o	0.13288	FTR H_o
Panchet versus Tenughat	2.34E−0.04*	Reject H_o	2.28E−04*	Reject H_o

FTR Failed to reject null hypothesis, * Significance level at 0.001

- The Land Use Land Cover (LULC) dynamics within the Maithon, Panchet, and Tenughat dam catchments reveal significant transformations, highlighting the influence of human activities and environmental shifts.
- The LULC transformations contributing most significantly to soil erosion and sediment yield in the Maithon, Panchet, and Tenughat dam catchments are the conversions of agricultural lands and forested areas to fallow lands and built-up areas.
- In all three catchments, bare land and fallow land consistently exhibit the highest soil erosion and sediment yield values, highlighting their vulnerability to surface runoff and sediment transport.
- Catchments dominated by mining activities in this region are highly susceptible to severe soil erosion and elevated sediment yield.
- In Tenughat dam catchment, consistent erosion, and improved connectivity from 1990 to 2020 indicate a steady supply of sediment delivery to the outlets. The areal expansion of erosion and connectivity is evident across all dams in 2020.
- The Konar and Panchet reservoirs exhibit the highest trapping efficiency at 93%, followed closely by Maithon at 92%. Tilaiya and Tenughat reservoirs have slightly lower trapping efficiencies at 88 and 90%, respectively.
- The high intensity of mining activities in the Tenughat Dam catchment substantially contributes to sediment production, resulting in a significant difference between the Tenughat and Panchet dams.
- Bare land and fallow land serve as the primary contributors to sediment yield across all three catchments.

7.12 Conclusion

In conclusion, the Damodar River basin (DRB) faces significant challenges due to land-use transformations, including declining agricultural lands, increased built-up areas, and deforestation. The Damodar River basin in India is highly regulated, with five major dams constructed to control disastrous floods in the region. However, the rich mineral resources in this catchment make it more vulnerable to land-use transformation and the intensification of soil erosion through intensive mining. Across all three catchments, a consistent trend of declining agricultural lands is observed, with significant portions transitioning to built-up areas and fallow lands. Improper land-use management and unplanned construction exacerbate sediment disconnection, reducing sediment delivery to downstream systems. Additionally, deforestation in upstream areas accelerates erosion, while agricultural practices on slopes disturb soil stability. These changes diminish reservoir storage capacity and reduce sediment trapping efficiency. Among the dams, Tilaiya Dam experiences the highest storage loss (36.01%), followed by Maithon (27.40%), Konar (25.43%), and Panchet (25.17%), with Tenughat showing the lowest storage loss (16.8%). Anthropogenic activities not only increase soil erosion intensity in the region but also contribute a substantial amount of sediment to the reservoirs. The Tenughat Dam catchment exhibits the maximum erosional intensity and is significantly different ($p < 0.001$) from the Maithon and Panchet dam catchments. The land-use classes in areas with very high soil erosion require sustainable management to conserve soil and reduce sediment yield to the main channel. Intensive mining and deforestation are among the primary causes of the excessive sediment input to the reservoirs (sinks). These dam catchments need sustainable land-use management to mitigate erosion and preserve reservoir efficiency.

References

1. Argaw M, Yohannes H (2024) Impact of land use/land cover changes on surface water and soil-sediment export in the urbanized Akaki River catchment, Awash Basin, Ethiopia. J Hydrol Reg Stud 52:101677. https://doi.org/10.1016/j.ejrh.2024.101677
2. Birhanu SY, Moges MA, Sinshaw BG et al (2022) Hydrological modeling, impact of land-use and land-cover change on hydrological process and sediment yield; case study in Jedeb and Chemoga watersheds. Energy Nexus 5:100051. https://doi.org/10.1016/j.nexus.2022.100051
3. Borrelli P, Robinson DA, Fleischer LR et al (2017) An assessment of the global impact of 21st century land use change on soil erosion. Nat Commun 8. https://doi.org/10.1038/s41467-017-02142-7
4. Brune GM (1953) Trap efficiency of reservoirs. Eos, Trans Am Geophys Union 34:407–418. https://doi.org/10.1029/TR034i003p00407
5. Calsamiglia A, Fortesa J, García-Comendador J et al (2018) Spatial patterns of sediment connectivity in terraced lands: anthropogenic controls of catchment sensitivity. L Degrad Dev 29:1198–1210. https://doi.org/10.1002/ldr.2840
6. Chuenchum P, Xu M, Tang W (2020) Predicted trends of soil erosion and sediment yield from future land use and climate change scenarios in the Lancang-Mekong River by using the modified RUSLE model. Int Soil Water Conserv Res 8:213–227. https://doi.org/10.1016/j.iswcr.2020.06.006
7. CWC (2001) Compendium on silting of reservoirs in India. Government of India, New Delhi, India: Reservoir Sedimentation Directorate, CWC
8. CWC (2020) Flood forecast monitoring
9. DVC (2021) Reservoir overview & salient features. In: Damodar Val. Corp. https://www.dvc.gov.in/cms-web/details-pages/113
10. DVC (2019) Sedimentation survey report of Maithon reservoir under NHP
11. DVC (2020) Sedimentation survey report of konar reservoir under NHP
12. DVC (2020) Sedimentation survey report of panchet reservoir under NHP
13. DVC (2019) Sedimentation survey report of tilaiya reservoir under NHP
14. ESRI (2024) Land use land cover map. https://livingatlas.arcgis.com/landcover/

15. Fryirs KA, Brierley GJ (2012) Geomorphic analysis of river systems. Wiley
16. Fryirs KA, Brierley GJ, Preston NJ, Spencer J (2007) Catchment-scale (dis)connectivity in sediment flux in the upper Hunter catchment, New South Wales, Australia. Geomorphology 84:297–316. https://doi.org/10.1016/j.geomorph.2006.01.044
17. Heinemarm HG (1981) A new sediment trap efficiency curve for small reservoirs. JAWRA J Am Water Resour Assoc 17:825–830. https://doi.org/10.1111/j.1752-1688.1981.tb01304.x
18. Iradukunda P, Bwambale E (2021) Reservoir sedimentation and its effect on storage capacity—a case study of Murera reservoir, Kenya. Cogent Eng 8. https://doi.org/10.1080/23311916.2021.1917329
19. John J, Rosamma CN, Thampi SG (2022) Assessment and prediction of soil erosion and its impact on the storage capacity of reservoirs in the Bharathapuzha River Basin, India. Environ Model Assess 27:77–103. https://doi.org/10.1007/s10666-021-09786-2
20. Kaffas K, Pisaturo GR, Premstaller G et al (2024) Event-based soil erosion and sediment yield modelling for calculating long-term reservoir sedimentation in the Alps. Hydrol Sci J 69:321–336. https://doi.org/10.1080/02626667.2024.2311756
21. Karan SK, Ghosh S, Samadder SR (2019) Identification of spatially distributed hotspots for soil loss and erosion potential in mining areas of Upper Damodar Basin—India. CATENA 182:104144. https://doi.org/10.1016/j.catena.2019.104144
22. Kothari CR (1990) Research methodology methods and techniques * we would. New Age International Publishers
23. Liang J, Chen C, Song Y et al (2023) Long-term mapping of land use and cover changes using Landsat images on the Google Earth Engine Cloud Platform in bay area—a case study of Hangzhou Bay, China. Sustain Horizons 7:100061. https://doi.org/10.1016/j.horiz.2023.100061
24. Liu Y, Jing Y, Han S (2023) Multi-scenario simulation of land use/land cover change and water yield evaluation coupled with the GMOP-PLUS-InVEST model: a case study of the Nansi Lake Basin in China. Ecol Indic 155:110926. https://doi.org/10.1016/j.ecolind.2023.110926
25. Ma T, Liu B, He L et al (2024) Response of soil erosion to vegetation and terrace changes in a small watershed on the Loess Plateau over the past 85 years. Geoderma 443:116837. https://doi.org/10.1016/j.geoderma.2024.116837
26. Máčka Z, Galia T, Škarpich V et al (2023) A method for assessment of sediment supply and transport hazard and risk in headwater catchments for management purposes. Environ Earth Sci 82:20. https://doi.org/10.1007/s12665-022-10707-z
27. Mhaske SN, Pathak K, Dash SS, Nayak DB (2021) Assessment and management of soil erosion in the hilltop mining dominated catchment using GIS integrated RUSLE model. J Environ Manage 294:112987. https://doi.org/10.1016/j.jenvman.2021.112987
28. Rahmani V, Kastens J, DeNoyelles F et al (2018) Examining storage capacity loss and sedimentation rate of large reservoirs in the Central U.S. Great Plains. Water 10:190. https://doi.org/10.3390/w10020190
29. Ren D, Liu S, Wu Y et al (2024) Quantifying natural and anthropogenic impacts on streamflow and sediment load reduction in the upper to middle Yellow River Basin. J Hydrol Reg Stud 53:101788. https://doi.org/10.1016/j.ejrh.2024.101788
30. Roy P, Roy A, Joshi P et al (2015) Development of decadal (1985–1995–2005) land use and land cover database for India. Remote Sens 7:2401–2430. https://doi.org/10.3390/rs70302401
31. Singh AK, Hasnain SI (1999) Environmental geochemistry of Damodar River basin, east coast of India. Environ Geol 37:124–136. https://doi.org/10.1007/s002540050369
32. Singh YK (2006) Fundamentals of research metholodgy and statistics, 1st edn. New Age International Publishers
33. Teshome DS, Moisa MB, Gemeda DO, You S (2022) Effect of land use-land cover change on soil erosion and sediment yield in Muger Sub-Basin, Upper Blue Nile Basin, Ethiopia. Land 11:2173. https://doi.org/10.3390/land11122173
34. Verma S, Kumar Verma R, Kant Tiwary R et al (2012) Relationships between land-use/land-cover patterns and surface water quality in Damodar River Basin, India. Glob J Appl Environ Sci 2:107–121

35. Verstraeten G, Poesen J (2000) Estimating trap efficiency of small reservoirs and ponds: methods and implications for the assessment of sediment yield. Prog Phys Geogr 24:219–251. https://doi.org/10.1191/030913300676742153
36. Verstraeten G, Poesen J, de Vente J, Koninckx X (2003) Sediment yield variability in Spain: a quantitative and semiqualitative analysis using reservoir sedimentation rates. Geomorphology 50:327–348. https://doi.org/10.1016/S0169-555X(02)00220-9
37. Wisser D, Frolking S, Hagen S, Bierkens MFP (2013) Beyond peak reservoir storage? A global estimate of declining water storage capacity in large reservoirs. Water Resour Res 49:5732–5739. https://doi.org/10.1002/wrcr.20452
38. Zafar Z, Zubair M, Zha Y et al (2024) Performance assessment of machine learning algorithms for mapping of land use/land cover using remote sensing data. Egypt J Remote Sens Sp Sci 27:216–226. https://doi.org/10.1016/j.ejrs.2024.03.003
39. Zeng C, He C, Liu Z et al (2022) Storage capacity of a karst groundwater reservoir associated with a large dam in a humid subtropical canyon karst area in southwestern China. Hydrogeol J 30:1989–2012. https://doi.org/10.1007/s10040-022-02551-z
40. Zhao Y, An R, Xiong N et al (2021) Spatio-temporal land-use/land-cover change dynamics in coastal plains in Hangzhou bay area, china from 2009 to 2020 using google earth engine. Land 10. https://doi.org/10.3390/land10111149

Chapter 8
Management Planning and Recommendation

Abstract This chapter outlines a comprehensive management plan and recommendations for addressing sedimentation issues in the Maithon, Panchet, Tenughat, and other reservoirs in the Damodar River Basin. The strategies focus on sustainable land use management, erosion control, and sediment retention to reduce the sediment load on reservoirs. Key recommendations include enhancing afforestation, reforestation, and social forestry, improving land use planning, and constructing sediment retention tanks, and check-dams in the upper catchment areas to reduce sediment flow. Sustainable mining practices and restrictions on illegal mining are emphasized to prevent soil erosion and sediment production. These integrated strategies aim to enhance sediment management, ensuring the long-term sustainability of water resources in the Damodar River Basin.

Keywords Retention tank · Sedimentation pond · Check dam · Dredging · Afforestation · Reforestation · Coal mining · Integrated watershed management plan · Site-specific management plan

8.1 Introduction

The Damodar River Basin (DRB) is a critical hydrological region in eastern India that houses the Maithon, Panchet, and Tenughat Dams which play pivotal roles in flood control, irrigation, and power generation. The upper catchment of the Damodar River Basin faces substantial challenges related to soil erosion and sediment yield, driven by extensive land-use transformations. Appropriate management planning is imperative to manage intensive soil erosion and sediment yield at the catchment level [1]. Various researchers across the globe focus on the topic of sustainable management for erosion and sedimentation [1–9]. The land use and land cover (LULC) dynamics and sediment yield variations in the Maithon, Panchet, and Tenughat dam catchments highlight critical environmental challenges requiring integrated management strategies. However, sedimentation poses a severe threat to their operational

S. A. Alam and R. Maiti, *Sediment Source to Sink: Deciphering Sediment Connectivity to Large Dams in Damodar River Basin*, Earth and Environmental Sciences Library,
https://doi.org/10.1007/978-3-031-87072-9_8

efficiency and longevity. Rapid deforestation, urbanization, and agricultural expansion have accelerated soil erosion, leading to increased sediment influx into reservoirs and reduced water storage capacity. Sediment accumulation not only reduces reservoir storage capacity but also impairs water quality and affects downstream ecosystems. Spatial analysis using geospatial tools and models like RUSLE, SWAT, IC, and SEH reveals erosion-prone zones and sediment pathways, identifying areas requiring immediate intervention. This chapter proposes a targeted management plan and actionable recommendations to address sedimentation challenges based on the comprehensive findings from the previous chapters.

8.2 Management Planning for Sedimentation Problems

Soil conservation focuses on reducing erosion in the catchment to ensure sustainable sediment management in both the channel and the reservoir. Sediment management within dams requires targeted interventions to mitigate reservoir siltation and maintain storage capacity. Installing sedimentation basins upstream serves as a proactive measure to trap incoming sediments before they enter the reservoir, thus minimizing deposition. Periodic desilting operations, such as dredging, are essential to restore lost capacity and ensure the long-term functionality of the reservoir. Additionally, sediment routing through the development of sluicing systems allows for controlled sediment passage during high-flow periods and maintains sediment equilibrium between the upstream and downstream areas. Integrated watershed management approaches involving land use regulation, community participation, and continuous monitoring using tools like SWAT modeling and remote sensing are crucial for long-term sediment control. Advanced technologies, such as real-time sediment tracking sensors and machine learning-based sediment yield prediction models, should be employed to support decision-making and proactive management. Coordinated efforts between policymakers, researchers, and dam authorities can harmonize land and water resource management to maintain reservoir efficiency, reduce environmental degradation, and ensure sustainable development in the Damodar River Basin. A conceptual strategic plan or flowchart for sedimentation management outlines the systematic steps and actions needed to address sediment buildup in reservoirs. It provides a clear framework for identifying problem areas, implementing conservation measures, and ensuring the long-term sustainability of the dam infrastructure (Fig. 8.1). A robust management plan to mitigate sedimentation in the DRB involves integrated strategies focusing on erosion control, sediment management, and dam operation optimization.

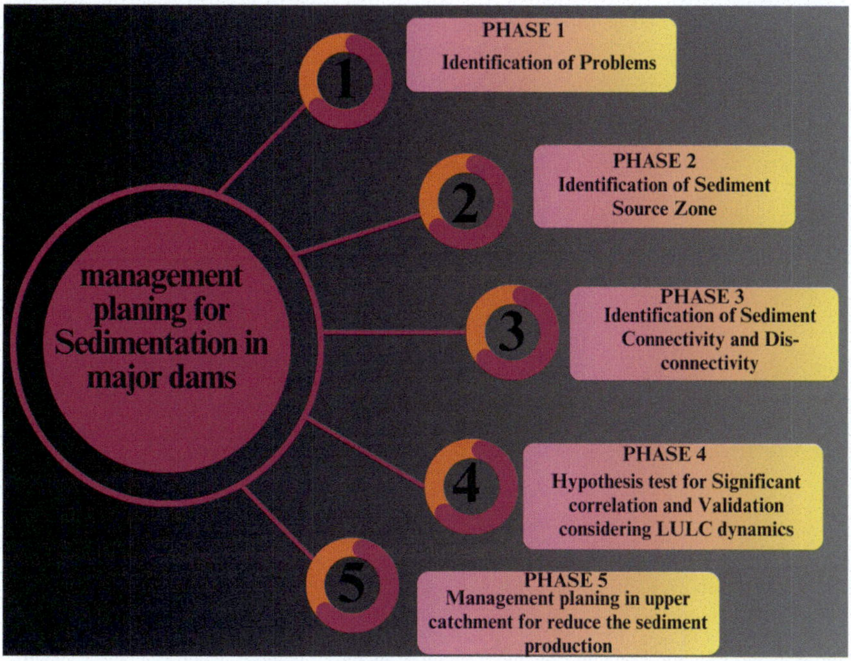

Fig. 8.1 A conceptual diagram illustrating the sequential process for identifying problem areas and their corresponding management plan

8.3 Recommendation

To effectively address sedimentation challenges in the Maithon, Panchet, and Tenughat Dams, the following recommendations are proposed:

- **Integrated Watershed Development Plans**: Integrated agricultural and Soil Management (ILSM) in micro watersheds, particularly upper catchments, seeks to alleviate soil degradation, improve water retention, and maximize agricultural productivity while preserving ecological balance (Fig. 8.2). The strategy combines sustainable land use practices, soil conservation, and watershed management measures to improve the catchment's health and minimize soil erosion.
- **Site-Specific Management Plan**: A site-specific management plan for soil erosion and sediment yield focuses on tailored strategies to address specific regions of erosion hotspots, fallow land area, and reduce sediment inflow into reservoirs. It incorporates localized interventions such as afforestation, contour farming, check dams, and sediment trapping measures to ensure sustainable watershed and reservoir management.

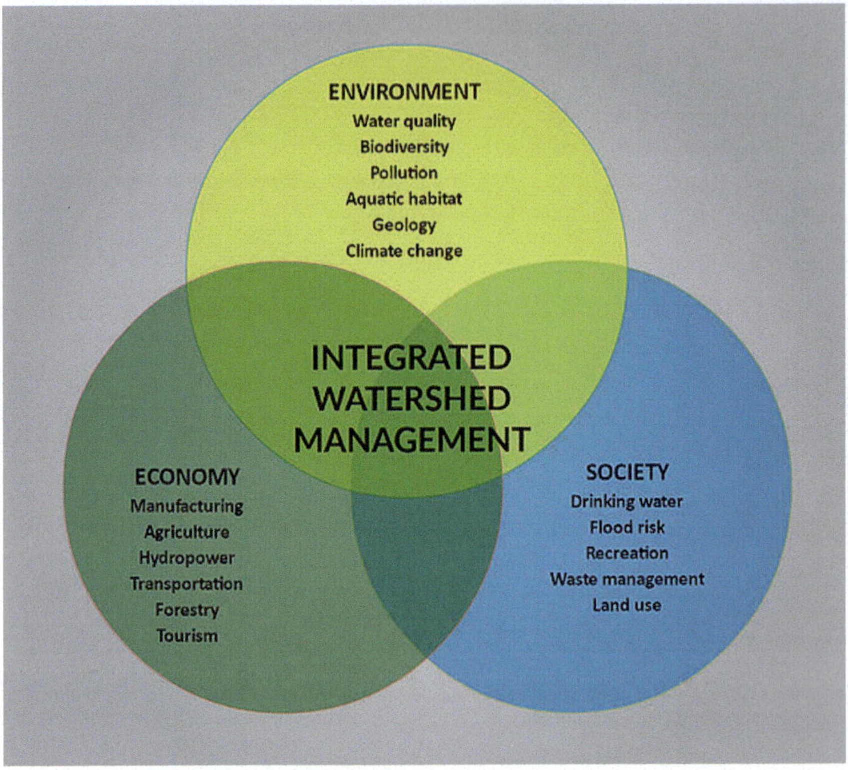

Fig. 8.2 An Integrated Watershed Management Plan (IWMP) aims to reduce soil erosion and promote socio-economic development in the dam catchment region (Source from: https://conservationontario.ca/policy-priorities/integrated-watershed-management, Date: 12.12.2024)

- **Erosion Control in Mining Areas**: Implement effective erosion control practices in open-cast mining areas including the use of cover crops, mulching, and other soil conservation techniques. Reducing open-cast mining and emphasizing underground mining can significantly mitigate environmental degradation, including soil erosion and habitat destruction.
- **Land use planning**: Land use planning plays a crucial role in managing soil erosion by promoting sustainable practices that balance economic activities with environmental conservation. Effective land use management significantly reduces erosion and sediment yield by implementing sustainable practices such as agroforestry, terracing, and controlled grazing. Establishing buffer zones along waterways and enhancing forest cover can stabilize soil, reduce runoff, and prevent soil erosion. Additionally, promoting soil conservation techniques like contour plowing and the use of cover crops can protect vulnerable areas from erosion.

- **Reforestation and Afforestation**: Reforestation involves replanting native vegetation in degraded or deforested areas to stabilize the soil, enhance water retention, and prevent erosion. Maximizing plantation on barren and fallow land is an effective strategy to combat land degradation, enhance soil stability, and reduce sediment yield. Afforestation, on the other hand, involves planting trees in barren lands and creating new forested areas that contribute to increasing vegetation cover, improving biodiversity, and reducing runoff. Both approaches play a vital role in mitigating sediment transport to reservoirs and improving overall watershed health.
- **Illegal coal mining**: Restricting illegal coal mining is essential to mitigate land subsidence and instability, which exacerbate land abandonment, and soil erosion and increase sediment production. These unregulated activities disrupt the structural integrity of the land, creating voids and collapses that accelerate soil erosion and sedimentation. Enforcing strict regulations, monitoring vulnerable areas, and promoting sustainable mining practices can minimize these impacts.
- **Check Dam**: Increasing the number of check dams in the upper catchment (Fig. 8.3) is essential for enhancing the in-channel retention of sediment. These structures reduce water flow velocity, allowing sediments to settle within the channel rather than being transported downstream. This not only mitigates reservoir sedimentation but also stabilizes streambanks, prevents soil erosion, and improves local groundwater recharge.
- **Dredging**: Dredging might not be an effective strategy for managing sediment in reservoirs with huge sediment loads due to its high cost, extensive logistical challenges, and limited capacity to address the continuous influx of sediments.
- **Retention tank or sedimentation pond**: The construction of additional water and sediment retention tanks in the upper catchment is crucial for effective sediment management. These tanks help trap and retain sediment-laden water, reducing the sediment load reaching downstream reservoirs. The construction of sedimentation ponds strategically in areas with high sediment yield can also significantly reduce downstream sediment transfer. These ponds can capture sediments before they reach the dams, preventing sedimentation and extending the life of the dams.
- **Social forestry**: Encouraging social forestry is a key strategy for the effective management of existing forests and for engaging local communities in sustainable resource management. By involving communities in planting, protecting, and utilizing forest resources, social forestry enhances forest cover, reduces soil erosion, and improves sediment retention.
- **Monitoring**: Monitoring and adaptive management are critical for effectively addressing sediment yield under climate extremes such as floods, cyclonic events, and intense rainfall. Continuous monitoring of sediment dynamics and hydrological changes enables the identification of emerging issues and the effectiveness of implemented strategies.

Fig. 8.3 A check dam has been constructed in the Rarha region of Jharkhand within the upper catchment of the Tenughat Dam (Source: Sk Asraful Alam, Date: 17.02.2024)

8.4 Conclusion

In conclusion, sedimentation in the Maithon, Panchet, and Tenughat Dams presents significant challenges to the long-term functionality and sustainability of these reservoirs. The integration of sediment management strategies, such as upstream sedimentation basins, desilting operations, and sediment sluicing systems, is essential to maintain reservoir capacity and mitigate environmental impacts. By adopting a multifaceted approach involving policy development, community participation, technological advancements, and operational strategies, sedimentation issues can be effectively managed. Moreover, ongoing research, real-time monitoring, and capacity-building initiatives will support adaptive management and ensure the resilience of these critical water resources in the Damodar River Basin. The implementation of these measures will not only enhance dam operations but also promote sustainable watershed management, ultimately safeguarding water quality, supporting local livelihoods, and contributing to the overall environmental health of the region.

References

1. Bhatti MT, Ashraf M, Anwar AA (2021) Soil erosion and sediment load management strategies for sustainable irrigation in arid regions. Sustainability 13:3547. https://doi.org/10.3390/su13063547
2. Blanco H, Lal R (2023) Soil conservation and management. Springer Nature Switzerland, Cham
3. Kassawmar T, Gessesse GD, Zeleke G, Subhatu A (2018) Assessing the soil erosion control efficiency of land management practices implemented through free community labor mobilization in Ethiopia. Int Soil Water Conserv Res 6:87–98. https://doi.org/10.1016/j.iswcr.2018.02.001
4. Morgan RPC (2009) Soil erosion & conservation. Wiley-Blackwell
5. Nasir Ahmad NSB, Mustafa FB, Muhammad Yusoff SY, Didams G (2020) A systematic review of soil erosion control practices on the agricultural land in Asia. Int Soil Water Conserv Res 8:103–115. https://doi.org/10.1016/j.iswcr.2020.04.001
6. Saljnikov E, Mueller L, Lavrishchev A, Eulenstein F (2022) Advances in understanding soil degradation. Springer International Publishing, Cham
7. Shit PK, Maiti R (2013) Management techniques of Rill-Gully erosion in Badland topography: experimental research at laboratory, plot and catchments scale. LAP LAMBERT Academic Publishing
8. Singh R (2023) Soil and water conservation structures design. Springer Nature Singapore, Singapore
9. Wang Z, Hu C (2009) Strategies for managing reservoir sedimentation. Int J Sediment Res 24:369–384. https://doi.org/10.1016/S1001-6279(10)60011-X

Chapter 9
Conclusion

9.1 Concluding Insights

The findings from this book provide a comprehensive understanding of sediment dynamics and connectivity in the Damodar River Basin, with a focus on mitigating sedimentation challenges in major reservoirs such as Maithon, Panchet, and Tenughat. By integrating advanced modeling techniques like SWAT, RUSLE, and SDR with geospatial analyses, the research highlights the critical relationships between soil erosion hotspots, sediment delivery ratios, and sediment connectivity indices over spatial and temporal scales. Key insights reveal that the upper and middle catchments, characterized by steep slopes, mining activities, and erodible soils, contribute significantly to sediment yield (SY) despite lower discharge levels. Sub-basins such as SB6, SB7, and SB12 emerge as critical zones of high sediment connectivity (SCI) and soil erosion, reflecting their vulnerability and importance for targeted interventions.

The Maithon, Panchet, and Tenughat dams have experienced an increase in sediment hotspot areas from 1990 to 2021, driven by the spatial expansion of hotspots and their merging with surrounding areas, despite a decline in their hotspot numbers. Tenughat exhibits the highest sediment yield among the major dams in the Damodar basin. Sediment connectivity is most pronounced in the upland and piedmont regions; however, dam structures significantly alter sediment dynamics. This results in a Basin Disconnectivity Index (BDI) value of 66.67, reflecting substantial disconnectivity across the entire basin. Positive correlations between SCI and SY further underline efficient sediment transport mechanisms, particularly in Piedmont and hilly regions. The Sediment Connectivity Index (SCI) and Basin Dis-connectivity Index (BDI) provide robust tools for quantifying and characterizing sediment dynamics, enabling a targeted approach to identify soil erosion hotspots and their impacts on dam sedimentation. The highest sediment trapping efficiencies of reservoirs like Konar (93%) and Maithon (92%) demonstrate their role in sediment retention, though challenges

S. A. Alam and R. Maiti, *Sediment Source to Sink: Deciphering Sediment Connectivity to Large Dams in Damodar River Basin*, Earth and Environmental Sciences Library,
https://doi.org/10.1007/978-3-031-87072-9_9

remain due to high sediment influx driven by land-use changes, including the conversion of forests and agricultural lands to fallow and built-up areas. Land use changes, particularly from agricultural and forested areas to bare and fallow lands, amplify soil erosion and sediment yield. Therefore, the book highlights the dynamic interplay between land use land cover (LULC) changes and sediment yield, emphasizing the need for sustainable land management practices to reduce sedimentation impacts.

Management recommendations derived from this research focus on reducing soil erosion through catchment-specific interventions, enhancing dam operational strategies, and integrating policy measures for sediment management. These findings have significant implications for sustainable reservoir management, ensuring the tenure of dam infrastructures and water resource availability in the region. The results presented in the book underscores the need for integrated sediment management strategies to mitigate erosion and optimize reservoir efficiency, particularly in erosion-prone catchments.

In conclusion, this work establishes a novel framework for sediment source-to-sink connectivity analysis, combining hydrological modeling, geospatial tools, and sediment management strategies. It serves as a valuable resource for researchers, policymakers, and practitioners, offering actionable insights into addressing sedimentation challenges in river basins worldwide. Future studies should focus on integrating climate change impacts and community-based management approaches to further enhance sediment management and water resource sustainability.

Index

S. A. Alam and R. Maiti, *Sediment Source to Sink: Deciphering Sediment Connectivity to Large Dams in Damodar River Basin*, Earth and Environmental Sciences Library,
https://doi.org/10.1007/978-3-031-87072-9

The manufacturer's authorised representative in the EU is Springer Nature Customer Service Centre GmbH, Europaplatz 3, 69115 Heidelberg, Germany. If you have any concerns regarding our products, please contact ProductSafety@springernature.com

Printed and bound by CPI Group (UK) Ltd, Croydon, CR0 4YY
01/10/2026
02225980-0011